12/24
#2

Astronomers' Observing Guides

For other titles published in this series, go to
www.springer.com/series/5338

Jeff Lashley

The Radio Sky and How to Observe It

with 125 Illustrations

Jeff Lashley
67 Sapcote Drive
Melton Mowbray
Leicestershire
LE13 1HG

Series Editor
Dr. Mike Inglis, BSc, MSc, Ph.D.
Fellow of the Royal Astronomical Society
Suffolk County Community College
New York, USA
inglism@sunysuffolk.edu

ISSN 1611-7360
ISBN 978-1-4419-0882-7 e-ISBN 978-1-4419-0883-4
DOI 10.1007/978-1-4419-0883-4
Springer New York Dordrecht Heidelberg London

Cover illustration: Details in radiation belts close to Jupiter are mapped from measurements that NASA's Cassini spacecraft made of radio emission from high-energy electrons moving at nearly the speed of light within the belts. The three views show the belts at different points in Jupiter's 10-hour rotation. A picture of Jupiter is superimposed to show the size of the belts relative to the planet. Cassini's radar instrument, operating in a listen-only mode, measured the strength of microwave radio emissions at a frequency of 13.8 gigahertz (13.8 billion cycles per second or 2.2-cm wavelength). The results indicate that the region near Jupiter is one of the harshest radiation environments in the Solar System. *Image Credit*: NASA/JPL

Springer is part of Springer Science+Business Media (www.springer.com)

To Irene and Christian, my greatest supporters

Preface

For most people, amateur astronomy means observing celestial objects in visible light, using the naked eye, binoculars, or an optical telescope. Even for many amateur astronomers electronics is a black art, a subject too difficult to comprehend.

Fear not – all is not that bad! This book does not assume you know anything about electronics. If you do, that's a bonus, and some parts of this book will not be new to you. Even if you have been involved in electronics or radio, probably radio astronomy is a new area for you. The emphasis throughout is about understanding how radio equipment works, what building blocks are needed, and about construction techniques. Once you grasp these initial basic concepts, it lays the foundation to taking your radio projects to the next level.

There is very little equipment commercially available for amateur radio astronomy. It always has been, and largely still is, the case that you will have to build at least some parts yourself, or often, modify some existing equipment to undertake radio astronomy. Two of the project radio telescopes in this book make use of surplus or low-cost satellite television components.

The book covers three broad areas. It begins with a non mathematical appreciation of astrophysics, particularly with regard to the radio frequency spectrum. The objects discussed are mostly limited to those that are of interest to amateur astronomers interested in studying radio sources. These include the Sun, Jupiter, meteors, and the Milky Way Galaxy. The Sun and Jupiter are the strongest of all the natural radio sources outside Earth. These are easy to detect with simple equipment of the type described here. With a little extra effort it will be possible to detect the noise background emitted from our Milky Way Galaxy, and even some of the discreet sources such as the Crab Nebula, Cassiopeia A, and Centaurus A.

The second broad topic is the theory of radio design, and an appreciation of the electronic components used. There are also sections describing the techniques you can use to build and test radio frequency circuits. Since the emphasis is on learning how radio telescopes work, there is a need to use some mathematics in order to design components. However, the formulae presented only require basic use of a pocket calculator. Many of the formulae only use addition, subtraction, multiplication, and division; some use squares, square roots, and logs. For example, instead of just telling you how to make a Yagi antenna, which works at 408 MHz, the chapter on aerial design provides the means you can use to build a Yagi for any practical frequency.

Following on from the theory of electronics and radios, the third section involves specific and detailed building projects. One of these projects – the VLF receiver – does not involve any electronic construction at all. Believe it or not you can use an ordinary computer sound card as a very low frequency radio receiver and plug a specially made aerial straight into the microphone port! The project chapters start with the simplest designs and work towards more complex receivers. If you are new to electronics, it is recommended you build the projects in the order presented. Your success at each stage will give you more confidence to continue on. As you work through the examples you will learn various aspects about how the systems work.

Our aim here is to cover the basics, provide you with some practical experience, and above all encourage you to learn more and become involved in the fascinating hobby of amateur radio astronomy. The explosion of advancement in modern electronics over the last twenty years has created many new opportunities for experimentation. It has never been easier than now to get involved and make your own radio systems. The micro chip era has made it possible to build powerful receivers in a small package with few components. As far as radio receivers go, radio telescopes are probably the simplest to design and construct. The goal of a radio telescope is simply to measure the amount of power received from an object at a fixed frequency using a radiometer, or how that power varies over a range of frequencies using a spectrometer.

Above all else, have fun and gain satisfaction is using instruments you build yourself.

National Space Centre,
Leicestershire, UK

Jeff Lashley

Acknowledgements

I must thank my colleagues in the BAA Radio Astronomy Group for their dedication and inspiration, particularly Lawrence Newell, Paul Hyde, Terry Ashton, Martyn Kinder, Andrew Lutley, David Farn, Norman Pomfret, Alan Melia, Mark Byrne, Tony Abbey, and John Cook.

My thanks go to Andy Smith for his kind permission in supplying radio meteor data.

Many thanks also to Alan Rogers and Madeleine Needles from the MIT Haystack Observatory for their support in assessing the innovative VSRT instrument.

Finally my thanks go to Christian Monstein for background information on the eCallisto project.

About the Author

Jeff Lashley is a technical support engineer at the National Space Centre in Leicester, UK. He has written regularly for Sunderland and Dundee newspapers. His most recent article on Radio Astronomy was published in the Radio Society of Great Britain magazine, Radcom.

Contents

Chapter 1

The Radio Sun

Our Sun is an ordinary star, and the nearest one to Earth. Its distance from us defines a convenient unit of measure in our Solar System, the astronomical unit, or AU. One AU is approximately 150 million km, which is the mean distance from Earth to the Sun. The solar diameter is 696,000 km, or approximately 0.7 million km, presenting an angular diameter of about 0.5° as seen from Earth's surface. Its spectral class is G2, and it lies centrally placed on the main sequence of the Hertzsprung–Russell diagram in Fig. 1.1.

Stars are composed of a mix of gases, mostly hydrogen and helium, far too hot in the center to contain solid matter. Hydrogen in the core fuses together to form helium, and in so doing it releases large amounts of energy. Gravitational forces act to compress the star into a volume with the minimum of surface area, a sphere. The dimension of a star is maintained as a stable balance between the internal pressure pushing out, caused by the high temperatures at the core due to the energy released by fusion reactions, and the gravitational force wanting to contract it. Stars maintain this equilibrium state for millions to even billions of years. The life expectancy of a star is closely related to its mass. The most massive stars live fast and die young, the smallest live a very long time.

Most ordinary stars, including the Sun, are not big producers of radio energy. The radio properties of our Sun are only significant to us due to its proximity. In order to understand the radio emission, it is first important to appreciate the structure and processes of the solar environment.

The Solar Core

The Sun is fluid object throughout; although it appears to have a clearly defined surface, known as the photosphere. This is by no means the limit of the solar environment. The photosphere merely represents that region where the gases of the Sun become opaque to radiation. Within the photosphere there are three distinct zones; working out from the center they are the core, the radiative layer, and the convective layer.

The core is where the thermonuclear reactions occur, the heart of the Sun, and where all the energy is produced. Its limit extends to 0.2 solar radii (R_0), although the bulk of energy production occurs within the central 10%of R_0.

J. Lashley, *The Radio Sky and How to Observe It*, Astronomers' Observing Guides,
DOI 10.1007/978-1-4419-0883-4_1,

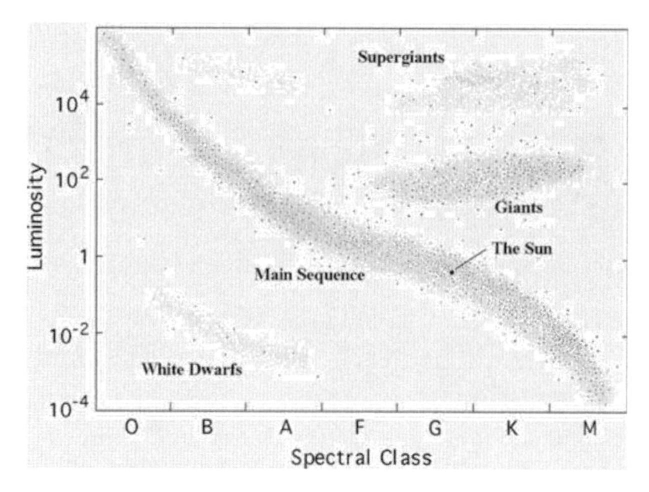

Fig. 1.1. The Hertzsprung-Russell Diagram. A plot of star colour against Luminosity. The vertical scale could be replaced by surface temperature.

In classical physics the fusion of four hydrogen nuclei to form a helium nucleus would require the particles to have sufficient kinetic energy (the energy due to their motion) to overcome the nuclear barrier forces. The kinetic energy of the atoms is largely a function of temperature, their energy increasing with higher temperatures. Hydrogen nuclei are fully ionized, which means they have no electron bound to them. Therefore they have a net positive charge. On the close approach of a pair of hydrogen nuclei they will repel each other most of the time and bounce off.

Although the core of the Sun has a temperature of 15 million K, this is insufficient to provide enough kinetic energy for the hydrogen to fully overcome the nuclear barrier forces and allow fusion. The temperature would have to be much higher. Fortunately, classical mechanics breaks down in this situation and is unable to explain nuclear fusion. Quantum mechanics can resolve this dilemma, however. Quantum mechanics is a theory that can explain many processes on the small scale of atomic structure, but fails to explain properties of the macro universe. These theories provide us with a better understanding of our universe than classical Newtonian mechanics ever could. Although quantum mechanics breaks down on the large-scale structure of the universe, and relativity breaks down at the small scales of atoms, scientists are still looking for the combined theory of everything! The process known as quantum tunneling allows a pair of atomic nuclei whose individual energies are lower than the nuclear potential energy barrier to fuse together. This is illustrated in Fig. 1.2.

The physicist de Broglie hypothesized that nuclear particles can exhibit wavelike properties. This was developed further by Schrödinger in his wave equation.

You can form a mental picture of a nuclear energy barrier by considering the rollercoaster ride pictured in Fig. 1.2.

If the car, without using an engine and ignoring frictional losses, was allowed to roll down the hill starting at point A, it would gain sufficient kinetic energy to ride over hill B but could not expect to reach a height higher than point C. If D represents the potential energy barrier between a pair of hydrogen nuclei, then it is very unlikely the pair will ever fuse unless the car tunnels through the barrier between C and E.

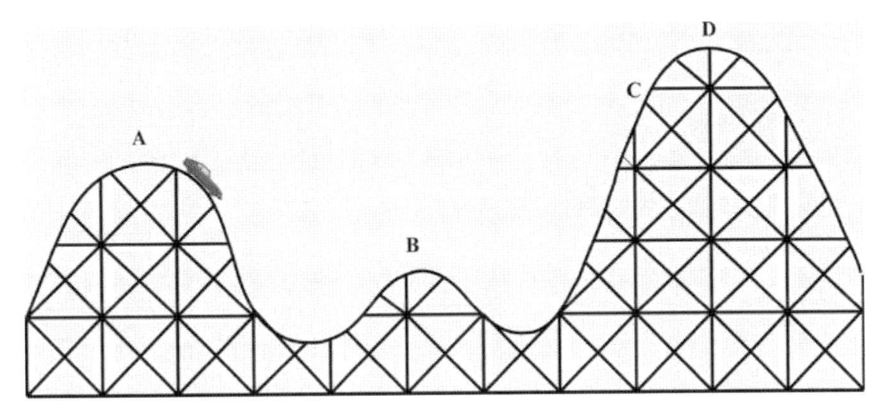

Fig. 1.2. Quantum tunneling.

The fusion process in the Sun is known as the proton-proton (PP) cycle. These PP chain reactions can take a number of routes to form helium, and which one it takes depends on the temperature of the stellar core. One of the reactions is more dominant than another for a given temperature. To take an example, the PP I chain reaction, which is dominant at temperatures from 10 million K to 14 million K, is given by these three formulae:

$$ {}_1H + {}_1H = {}_1D + e^{+} + v_e $$

$$ {}_1D + {}_1H = {}_2He^3 + \gamma $$

$$ {}_2He^3 + {}_2He^3 = {}_2He^4 + 2{}_1H $$

The leading subscript refers to the number of protons in the nuclei. Note in the first reaction that one of the protons is converted into a neutron, with a consequent loss of mass. This mass is converted into energy. The positron generated here immediately combines with an electron, and the resulting annihilation generates a gamma ray photon.

In the second reaction one helium 3 isotope and a gamma ray photon is produced. In the final stage two helium 3 isotopes combine to form stable helium 4 and two protons.

The first reaction is very slow to occur, because it relies on the rare quantum tunneling process. Once this is complete the rest of the reaction occurs more quickly.

The bulk of energy production in the core of the Sun is in the form of gamma ray photons, which pass into the next layer.

The Radiative Zone

The radiative zone acts like an insulator helping to maintain the high core temperatures of 15 million °C. Here there is no major mixing of the plasma gases. The gases in the radiative zone are so opaque to radiation that the gamma ray photons are absorbed and re-emitted many times over. The average path length for a given photon before interaction is about 1 mm – this is very opaque! It can take the photon 50,000 years or more to emerge.

The Convective Zone

Finally the energy enters the convective zone. Here energy transport is achieved by warm gases rising, and cool gases falling in a cyclic pattern.

The hot gases rise towards the photosphere, cooling near the surface as energy is once again radiated from the photosphere. Rising cells of gas are visible on the photosphere surface as a granular pattern that is familiar on high resolution photographs of the Sun.

The Photosphere

The photosphere is the surface that is seen visually in optical telescopes. It is impossible to directly observe the solar interior below the surface. One of the ways astronomers can study the interior is by helioseismology. Like seismology on Earth, vibrations and oscillations in the solar surface can be observed. The telltale ripples and vibration frequencies give a lot of information about the structure of the interior layers.

It is on the photosphere that we see sunspots. These are active regions that appear dark by virtue of their cooler temperature. The black appearance is only relative to the brighter surroundings; the spot temperatures are still around 4,000 K about 2,000 K below that of the average surroundings.

Sunspots and the activity level of the Sun undergo cycles of approximately 11 years duration. At the peak of activity spot numbers reach a maximum, and when the Sun is quiet there may be long periods when no spots are seen at all.

Sunspots are a visible indication of magnetic anomalies on the surface. In spot groups the polarity of neighboring spots is often opposite. This gives rise to magnetic loops, which can enable matter to stream along the loop and back down to the surface. There are exceptions, however, where the second pole is so diffused that a single unipolar spot can exist.

Sunspots develop slowly from small pore-like structures into large spots and groups of spots. The seed pores occur on the boundaries of the granulation cells, which are manifestations of the upwelling of warm gases. Most pores simply disappear, but a few develop into large activity regions that can ultimately generate huge bursts of energy in the form of solar flares. As we shall see flares are of significant interest to the radio astronomer.

The Magnetic Dynamo

There is still much to learn to fully understand how stars and planets form magnetic fields. The mechanism by which a magnetic field is maintained is known as the dynamo process. The dynamo is a self-exciting system, where the motions of electrically charged particles flow in a conductive medium, producing circulating currents that interact with the existing field to maintain the magnetic field strength.

As mentioned previously the Sun is a fluid object that exhibits differential rotation. This means the equatorial regions rotate more quickly than the polar zones. It is thought that over the course of a solar cycle, the magnetic field gets twisted and disturbed, inducing the active regions we see. This eventually builds to a peak over the 11-year cycle and begins to settle again. This 11-year cycle is an average figure; it can be a little shorter or a little longer, but after each maximum occurs, the magnetic polarity of the poles invert. In a sense the full cycle of the sun has a mean period of 22 years.

The Chromosphere

The layer above the photosphere is the chromosphere. This is the beginning of the solar atmosphere. The matter residing in the chromosphere and the layers above it are no longer opaque to light. In this region between 5,000 and 30,000 spicules are observed at any one time. Rising to a height of 10,000 km spicules are streams of matter flowing outward at around 20 km/s, rising to a temperature of 10,000 K, hotter in fact than the photosphere. The sheer number and velocity of spicules suggest a mass flux twice that of the solar wind. By processes still not well understood a proportion of the matter streams return to the chromosphere.

The Corona

The corona is a vast rarefied zone around the Sun. There is no well-defined edge to it. In fact matter escapes the solar environment all the time in the form of the solar wind, which extends to the boundaries of the Solar System and no doubt beyond.

The temperature of the corona in the close proximity of the Sun is around 1–2 million K at quiet times, but can reach several million K above active regions. The temperature is far from uniform at any time. At these very high temperatures the matter emits a broad range of thermal electromagnetic radiation, but it is particularly interesting in the X-ray spectrum, where X-ray bright zones indicate the presence of underlying active regions. Although light is also radiated from the corona, it is weak in comparison with the emission from the photosphere, so it only becomes visible to the observer during total eclipses. All of the radio emissions from the Sun originate in the chromosphere and corona, as we shall see. So radio astronomy helps us probe and study the processes and structure of the solar atmosphere.

The question of why the corona is so hot and what heats it up has been a problem for scientists for years. It has led to many theories and is still a controversial subject. Recent observations made by James Klimchuk using instruments on board the NASA-funded XRT X ray telescope and the ultraviolet instrument EIS on board the Japanese satellite Hinode have been used to test a model that nano-flares are the prime source of coronal heating. Klimchuk investigated super hot plasmas of between 5 and 10 million K, which can only be accounted for by the bursting effect of flare activity. These super hot plasmas cool very quickly, passing on energy to their surroundings and giving rise to a general 1 million K background. We know that large flare outbursts cause significant coronal heating above them, but the relative lack of large flares for a large part of the solar cycle does not see the coronal temperature falling much below 1 million K even then. The breakthrough made by Klimchuk shows evidence of a constant rate of miniature flares occurring all the time, which is capable of maintaining the minimum temperature of the corona throughout the solar cycle.

The Quiet Sun and the Blackbody

The concept of the blackbody is fundamental to radio astronomy. It enables us to estimate the temperature of objects by observing the objects at only one wavelength of electromagnetic radiation. A blackbody is defined as a perfect absorber of radiation, as well as a perfect radiator. In other words, all radiation impinging on a blackbody will be absorbed, with no reflection or scattering; it is totally opaque. This is the theoretical ideal, of course, and nothing is perfect, but stars are good approximations of blackbodies. Despite the name, a blackbody is not always dark! After all, it has to be perfect radiator, too. Blackbodies emit a characteristic energy spectrum, defined by Planck's equation, where the peak output wavelength depends on the temperature of the blackbody. By measuring the received intensity at any wavelength, by a process of curve fitting, the whole blackbody spectrum can be derived.

Cool stars exhibit peak radiation in the infrared spectrum; hotter stars peak in the visible spectrum, and very hot objects peak in the ultraviolet. The Sun's surface temperature is approximately 6,000 K, resulting in a peak blackbody radiation wavelength in the yellow region of the visible spectrum. Hence we refer to our Sun as a yellow star. Figure 1.3 illustrates the way stellar intensity is related to its spectrum and is a plot of the Planck's equation.

By measuring a spectrum or just part of the spectrum, we can infer its temperature by fitting the data to a Planck curve. The same curve can be expressed as a graph of intensity against wavelength, or it can be calibrated as temperature against wavelength. This method is used in radio astronomy all the time to estimate an object's temperature. However the results need to be treated with extreme caution. By observing solar flare radio bursts at, say, 20 MHz, a temperature could be derived. This is almost certainly going to have a value of tens of thousands or even millions of K. By fitting this data to a Planck curve, the estimated peak output would suggest a temperature many times that which is accepted for the Sun. The quiet Sun has an immeasurably small output in the HF radio spectrum. Unfortunately the active Sun significantly deviates from the ideal blackbody spectrum in the radio region.

The answer to this problem lies in the fact that a solar flare energy output is not produced by thermal processes, so calculating a temperature from such data yields inaccurate results. Yet radio astronomers still report such results as a temperature, known as the equivalent temperature. Despite the fact that the temperature value

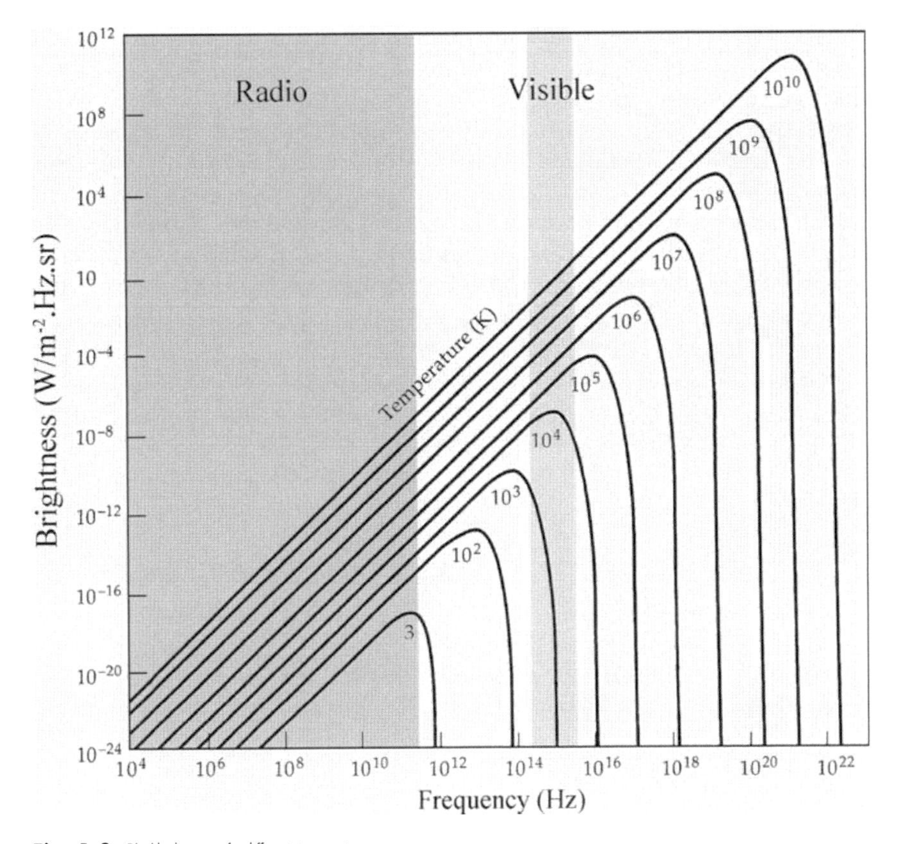

Fig. 1.3. Blackbody curves for different temperatures.

has no physical meaning, it is a useful figure that can be directly compared with other results, providing a measure of the strength of a radio signal. The equivalent temperature is therefore only valid for the frequency at which the measurement was taken, and the frequency must be quoted along with this temperature.

Solar Flares

Classification of Flares

Solar flares are classified by their size and duration, by their morphology, and by their magnetic topology. Flares are very complex phenomena, which can occur in white light, or they may be more restricted in their spectral output.

There are two basic types of flare, impulsive and gradual. Impulsive flares are of short duration, in the order of seconds to minutes. Many impulsive flares are fully contained within the Sun, although some are thought to induce mass ejection.

Gradual flares can last many minutes or even hours. This category of flare can generate huge energy output and eject matter that escapes the Sun. Fully developed flare events progress from the impulsive into the gradual phase.

Optical classification, referred to as the Hα importance, uses the scale S (sub flare sometimes also class 0), 1, 2, 3, 4, according to the area of the solar disk involved (see Table 1.1).

Today flares are routinely detected by satellite, usually in the X-ray band. The classification scheme uses one of the following prefixes – A, B, C, M, or X – X being the strongest class. The class is further subdivided and given a number such as C3, the number being an integer in the range 0–9 except for the X class, which can be any integer. The unit of measurement is W/m^2 (see Table 1.2).

For example a class C5 flare would have a strength of 5×10^{-6} W/m^2.

Radio outbursts at centimeter or meter wavelengths are classified on a scale of I–V in Roman numerals. See Table 1.3. This scheme was devised in the early days of radio astronomy, before the nature of flares was understood. The order of the numbering is relatively meaningless in the physical world.

Type III and V are associated with the impulsive flares or the impulsive phase of eruptive flares. They are generated by the acceleration of electrons along magnetic field lines into the corona.

Table 1.1. Optical classification scale for solar flares known as H α importance

Importance	Area (A) in square degrees
S	$A < 2$
1	$2.1 \leq A < 5.1$
2	$5.1 \leq A < 12.4$
3	$12.5 \leq A < 24.7$
4	$A > 24.8$

Table 1.2. Satellite based classification of solar flares based on X-Ray emission

Importance	Strength (W/m^2)
A	10^{-8}
B	10^{-7}
C	10^{-6}
M	10^{-5}
X	10^{-4}

Table 1.3. Solar radio burst classification scheme

Class	Duration	Bandwidth (MHz)	Drift rate frequency
Type I	Seconds	5	
Type II	Minutes	50	20 MHz/min
Type III	Seconds	100	20 MHz/s
Type IV	Hours	Wideband continuum	
Type V	Minutes	Wideband continuum	

Type II and IV are associated with the eruptive flares and coronal mass ejections (CME). The Type III are thought to be generated by flare-induced shock waves in the corona traveling at around 500 km/s. The Type IV is thought to be produced by magnetic reconnection following a coronal mass ejection.

How Solar Flares Form

Flares occur when charged particles are suddenly accelerated. The energy to induce the acceleration must come from the magnetic fields surrounding the active areas.

We saw earlier that groups of sunspots can be made up of pairs of spots with opposite magnetic polarity, between which there are looped magnetic field lines. Imagine a boundary line between these two spots we refer to as the neutral line. If the magnetic loop is perpendicular to the neutral line it is in the potential configuration. As the spots develop and move they can slide relative to each other along the neutral line. When this happens the field loops are no longer perpendicular to the neutral line. If this process continues to extreme levels, the field becomes nearly parallel to the neutral line, known as a sheared magnetic field. These sheared fields have more magnetic energy than the potential configuration. At this point magnetic instabilities can occur, breaking field lines and releasing energy, returning the magnetic field to the potential configuration by the process known as magnetic reconnection. The release of the free magnetic energy provides both thermal and non thermal energy to the surrounding plasma, accelerating the charged particles.

During the impulsive phase there is a rapid increase in intensity of radiation across the electromagnetic spectrum, particularly in the hard X-ray region, extreme ultraviolet, and the decimetric and centimetric radio wavelengths. These events originate in magnetic loops of the chromosphere, in the highly non-potential fields described above, where the electron densities exceed 10^{13} cm^{-3}.

The accelerated particles stream along the field loops and lose energy by electron-electron collisions and electron-ion interactions to produce the hard X-ray photons. In the low density parts of the loop ($<10^{10}$ particles per cubic meter) only a small proportion of the energy is lost, but when particles encounter the higher density regions the rate of energy loss significantly increases, stopping the electrons and heating the plasma.

Most flares are confined, impulsive types, which merely cool in the main phase by conduction into the cooler chromosphere, or by radiation. Conductive cooling is greater for long loops, although radiative cooling is lower in higher density loops. So short, dense flares dissipate first and long, lower density loops last longer.

The more powerful eruptive flares continue to emit radiation in the main phase, sometimes for several hours. It is in these eruptive flare events that the magnetic loops first break and then begin to reconnect again, releasing the free magnetic energy and heating the plasma to temperatures as high as 20 million K. Magnetic reconnection occurs quickly in the lower levels, but much more slowly at greater heights, leading to the long-duration events that we observe.

The loop heating generated by the reconnection events make the loops visible and usually appear as a pair of bright loops.

Solar Flare Radio Bursts

The Type III fast drift bursts occur whenever there are active regions present on the Sun. At decimeter wavelengths Type III bursts drift quickly from high to low frequencies, as a jet of electrons is ejected upward into lower density corona layers and about 90% of Type IIIdm bursts drift from low to high radio frequency, suggesting at times there are also jets of electrons descending into to denser regions. About half of the IIIdm bursts are associated with hard X-ray emission, suggesting a strong correlation with the impulsive phase of fully developed flares. Following the Type III bursts are often extended duration Type V continuum emissions during the main phase of the flare.

More than 90% of all Type II bursts are produced by solar flares, although not all flares generate Type II bursts. They are rarely seen in smaller flares with an optical classification less than two and occur in about one third of the flares with a Hα importance of between two and three. However they are often associated with small flares simply because small flares are far more common. About 70% of Type II events occur along with a coronal mass ejection; it is still unknown whether it is the CME or the flare blast wave that is the cause of the burst. Type IV continuum emission usually occurs alongside Type II bursts.

What, you might ask, is a Type I burst associated with? The original classification of radio flares dates back to 1963 and is based only on the morphology of the radio spectral profile, not on the physics behind the events. Accompanying Type IV continuum emission is often short, sharp, spiky noise usually classified as Type I.

At this time, not associated with flare activity, the slowly varying component of solar radio emission is prominent in the 3–60 cm wavelength band. This emission strongly follows the 11-year sunspot cycle, and not surprisingly it is associated with sunspot groups. The emission occurs in radio plages, up to 100,000 km above the active regions. High-resolution radio mapping has shown the central parts of approximately sunspot-size largely circular polarized radio energy emissions, while the surrounding plage has a more random polarization.

The Quiet Sun

At short wavelengths, those less than 1 cm, a high resolution radio telescope would show the solar disk subtends and angle of about half a degree, the same as seen visually. However at wavelengths of around 10 cm, the disk appears slightly larger, with a significant limb brightening especially in the east-west direction. While limb darkening in the optical is well known, and caused by increasingly cooler layers of gas towards the photosphere surface, limb brightening in radio is caused by the increasing temperature of the gases in the chromosphere. Observing at longer wavelengths the Sun subtends an angle of ever increasing diameter (Fig. 1.4).

From these observations it's clear that short wavelength radiation reaches us from closer to the solar surface and increasingly longer wavelengths relate to higher altitudes. The frequency of detected radiation comes from just above what is known as the critical layer for that frequency and is directly dependent on the electron density. As one would expect, the electron density decreases with

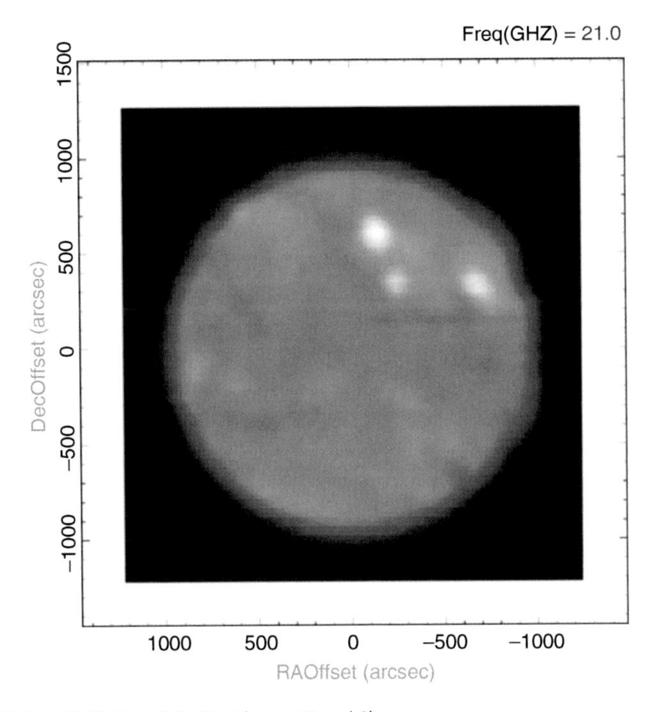

Fig. 1.4. The Sun at 21 GHz taken with the 37 m telescope at Haystack Observatory.

increasing altitude and offers astronomers a way of estimating electron densities for varying solar altitudes.

Non-outburst radio emission in the millimeter wavelength range occurs as a consequence of Bremsstrahlung thermal emission. Bremsstrahlung (breaking radiation) is caused by thermal electrons being deflected in the presence of the electric fields of ions in the chromosphere (Fig. 1.5).

At longer wavelengths of a few centimeters, gyroresonant processes are more important, whereby electrons are accelerated into spiral paths traveling along magnetic field lines. For long wavelengths, thermal emission is insignificant, and the bursting outputs of non thermal processes dominate.

The Solar Wind

The solar wind is a constant outflow of particles from the Sun. These give rise to radio bursts caused by instabilities when the solar wind encounters shock wave fronts passing through the Solar System. The shock wave fronts in turn originate in the corona of the Sun and provide us with a remote method of observing such irregularities in the Solar System, out as far as the planet Jupiter. However the frequency of the bursts is very low and is sadly not observable from the Earth's surface due to the ionospheric cut off.

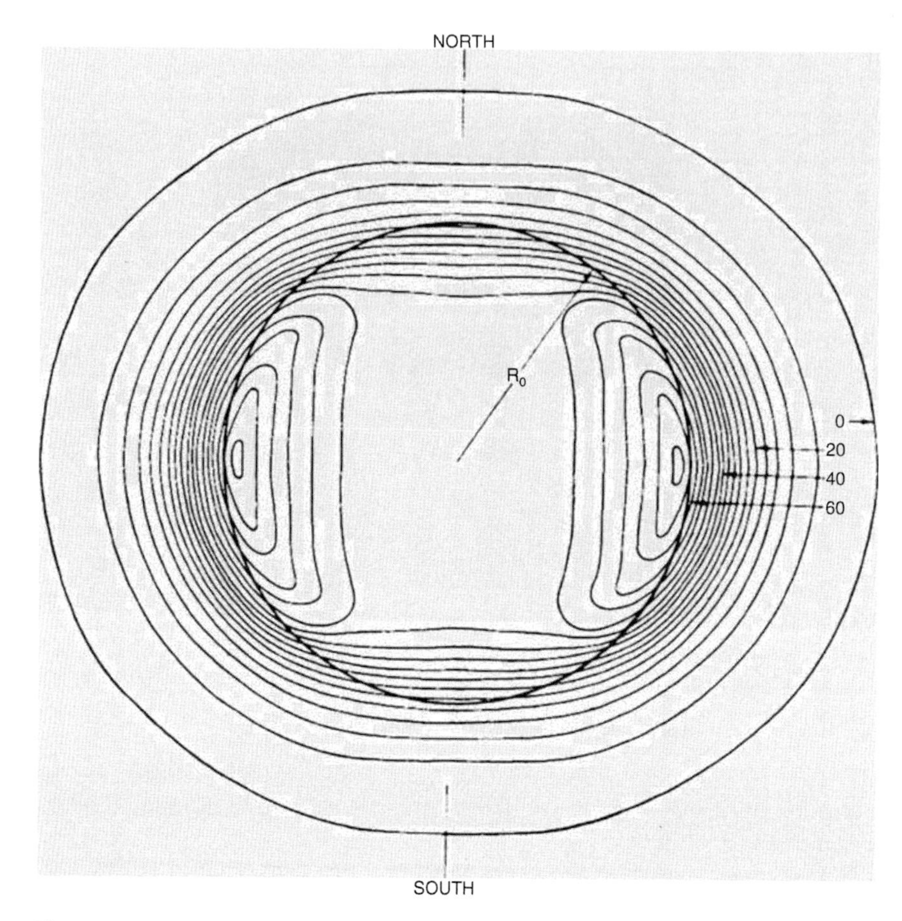

Fig. 1.5. A contour map of the Sun at 1.4 GHz. The circle of radius R_0 is the diameter of the visible Sun. (Based on data from W.N. Christiansen, and J.A. Warburton, *Australian Journal of Physics*, 1955).

Type II and III bursts are usually the only ones that can be tracked for any length of time by spacecraft into the deeper Solar System. They can be tracked for up to several hours, or maybe even for more than a day.

Interplanetary radio bursts occur at or near to the local plasma frequency. Hence they are dependent upon the electron density at the source. Once again the electron density continues to decrease with increasing distance from the Sun; therefore the critical frequency of emission decreases with distance, to the order of less than 100 kHz at about 50 AU from the Sun.

Type II bursts are seen usually only faintly at great distance from the Sun. The shock waves excite electrons to around 10 keV to produce Langmuir waves, which are converted to radio energy by a similar process as that of Type III radiation. This is particularly useful to Earth scientists tracking shock waves approaching Earth. Type III bursts are more energetic, and resulting sub relativistic jets of electrons with energy levels up to 100 keV are more commonly observed and better understood.

Amateur astronomers cannot directly observe the solar wind and its radio properties. However there are a couple of indirect means we can use to study its effects. Firstly there is the variation of long distance radio propagation of communication signals, and secondly we can check out the effects on local magnetic field.

In the project section of this book the VLF receiver is one instrument designed to monitor propagation effects, albeit in that case it's more sensitive to ionospheric X-ray exposure. It has long been known that HF radio propagation is strongly influenced by the solar cycle. While we will not be exploring HF propagation in this book in any detail, it would constitute an interesting long-term addition to a radio solar observing program to monitor the received signal strength of amateur radio beacons. The reader is strongly encouraged to get involved in amateur radio, and even progress to obtaining a transmitting license.

The use of a sensitive magnetometer can be part of the observer's arsenal of tools in order to monitor the solar influence on Earth. Charged particles of the solar wind that are captured by Earth in the magnetic polar regions not only create aurorae but also strongly influence the ambient magnetic field. Although the construction of a magnetometer is not covered in this book, the UKRAA arm of the British Astronomical Association's Radio Astronomy Group is shortly going to offer an instrument for just such observing programs.

Chapter 2

Jupiter

Jupiter is one of the most interesting Solar System objects to observe, whether that's with an optical or radio telescope, as well as one of the easiest. Jovian radio emission is very strong, at times (at HF radio frequencies) rivaling if not surpassing the Sun in signal strength.

The Structure of Jupiter

Spacecraft studies of small gravitational effects on spacecraft trajectory suggest that Jupiter has a small solid core of around 20 times that of Earth's mass, and a radius of around $0.2R_j$ where R_j is the Jovian planetary radius of 69,084 km (Fig. 2.1).

The layer surrounding the core is highly compressed liquid metallic hydrogen, which extends to a distance of $0.78R_j$. Above this lies a still dense atmosphere of mostly molecular hydrogen.

How can hydrogen be metallic, you may ask? Natural hydrogen atoms consist of a proton orbited by a single electron. Metallic hydrogen is formed when the atoms are compressed at very high pressures. This pressure is enough to reduce the average distance between hydrogen nuclei to less than the Bohr radius (the smallest orbital diameter for the electron). At these pressures the electrons become unbound from the protons and are free to move around, making the hydrogen electrically conductive, like a metal. At even more extreme pressures hydrogen would crystallize into a solid metallic lattice, although this is not the case for Jupiter, as indicated by the amount of polar flattening due to rotation.

The fluid conductive hydrogen layer plays an important role in generating the dynamo effect and the strong magnetic field we observe around the planet. The magnetic field plays an important role in the generation of radio emissions.

Jovian Magnetic Field

The dynamo effect in planets is similar to the dynamo effect in stars we saw earlier and is again a self-exciting system, which means it functions without any external input. Planetary dynamos are still not well understood, but to get some idea of how we think they work consider Fig. 2.2.

J. Lashley, *The Radio Sky and How to Observe It*, Astronomers' Observing Guides,
DOI 10.1007/978-1-4419-0883-4_2, © Springer Science+Business Media, LLC 2010

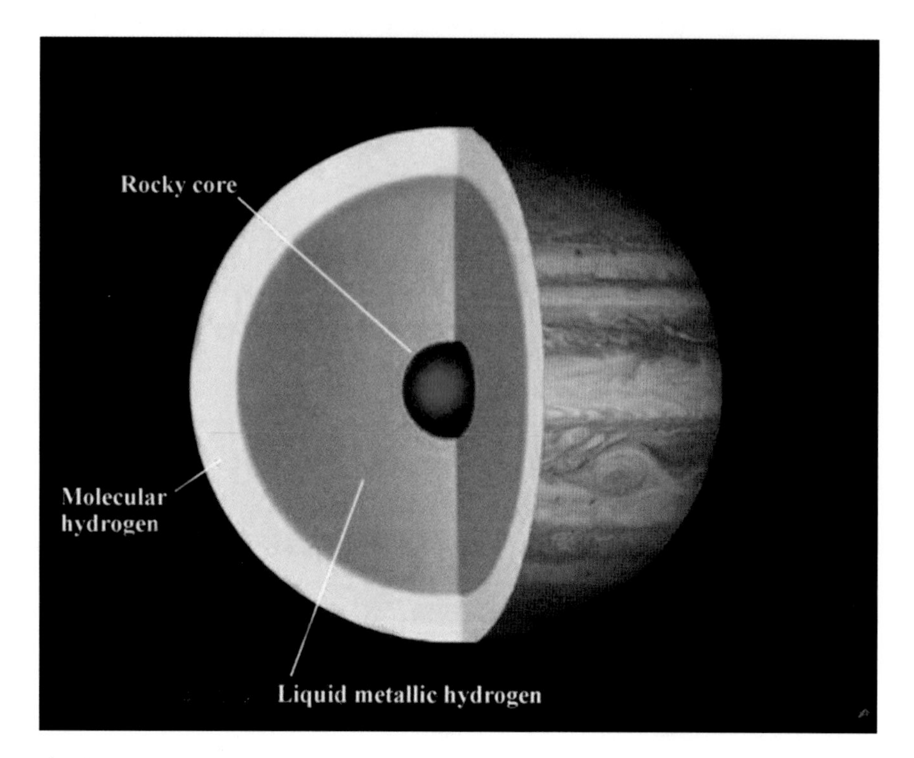

Fig. 2.1. The internal structure of Jupiter.

The conductive disk is rotating about an axis. A current loop is formed between the disk and the axis, the loop being completed by conduction through the disk itself. The current loop generates magnetic flux, which is intercepted by the disk generating more current. The conductivity of the disk is not perfect; in fact it is relatively poor, and hence it has considerable resistance. The current growth in the Jovian dynamo is counteracted by retarding Lorenz forces acting on the disk. In this way the system settles into an equilibrium state.

The Jovian Magnetosphere

From the earliest days of the detection of radio waves from the planet Jupiter in the 1950s, it was known there must be a significant magnetic field surrounding Jupiter. The bursting nature and strong output of radio energy in the HF spectrum, up to a sharp cut off at almost 40 MHz, equated to a temperature of about 3×10^{15} K, assuming it was due to blackbody radiation and was evenly distributed across the Jovian surface. Visible evidence alone clearly shows Jupiter cannot be that hot. The radio-generating process must be of non thermal origin. In order to begin to understand the radio output of Jupiter we need to build up a picture of the Jovian magnetosphere.

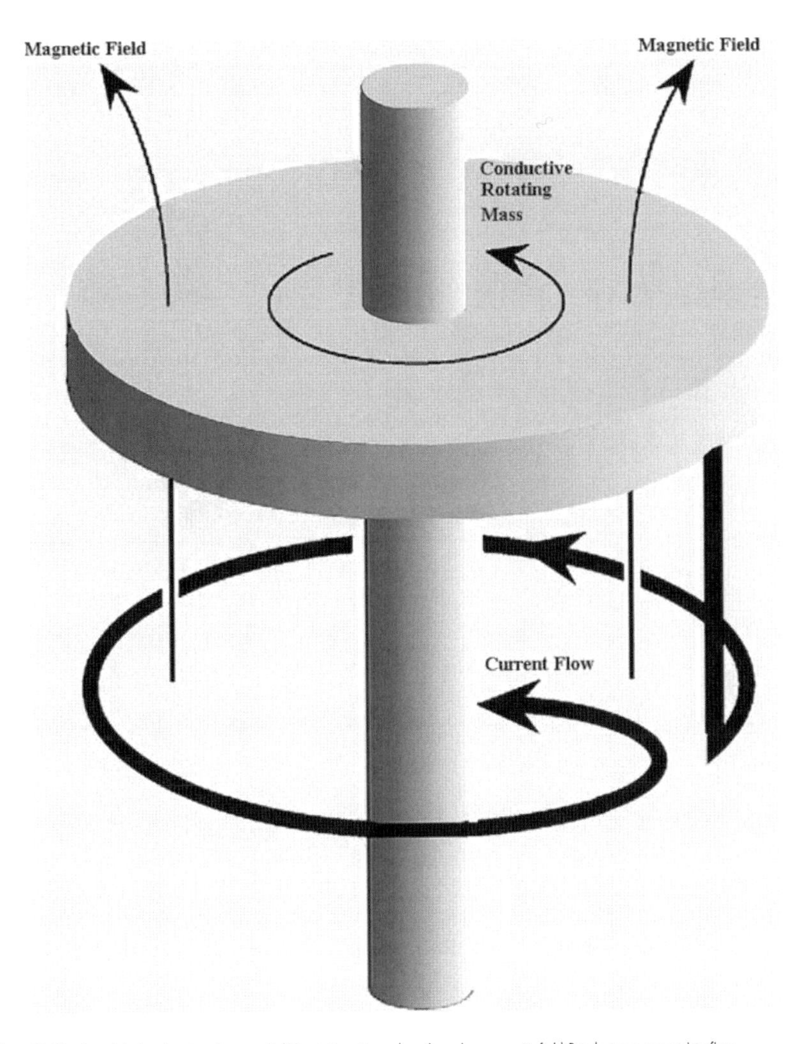

Fig. 2.2. A model of a planetary dynamo. A disk rotating at speed ω through a magnetic field B inducing a current I to flow.

The magnetosphere is the envelope surrounding Jupiter where the magnetic fields influence the motion of charged particles. It is both extensive and complex. It is convenient to discuss the Jovian magnetosphere in three parts: the inner, middle, and outer magnetosphere. This is because the dominant processes are different in each zone.

The Inner Magnetosphere

This region extends out to about six Jovian radii and is controlled by the predominantly dipole magnetic field generated within Jupiter. Studies carried out by *Pioneer 10* and *11* and *Voyager 1* and *2* show there is a significant asymmetry in the magnetic field, where the north pole has a magnetic field strength of 14 Gauss, and the south pole only 10.4 Gauss. The field is a dipole (a simple two-pole system,

Table 2.1. The longitude coordinate of Jupiter is broken into three systems, each based on a slightly different rotation period

System	Relevance	Rotation period
I	Cloud tops within $\pm 10°$ of equator	$9^h\ 50^m\ 30^s$
II	Cloud tops north or south of the 10° latitudes	$9^h\ 55^m\ 40.6^s$
III	The magnetosphere	$9^h\ 55^m\ 29.7^s$

like a bar magnet), which is tilted with respect to the planet's rotational axis by approximately 9.6°, toward a system III longitude of 202°. You may be familiar with the System I and II coordinate systems used for defining the position of cloud features on the visible surface of Jupiter. The system III was developed for use in radio astronomy, as a more accurate means of monitoring the motion of features in the magnetosphere. The different systems are detailed in Table 2.1.

The inner magnetosphere consists of a cloud of trapped charged particles. To understand why the particles are trapped, and don't simply get absorbed into the atmosphere, consider the motion of an electron inside the field. If an electron is traveling perpendicular to the field lines, a force known as the Lorenz force is induced by the magnetic field and deflects the electron into a circular orbit. The radius of the orbit is proportional to the velocity of the electron, and inversely proportional to the magnetic flux density. The rate of rotation around the magnetic field line is known as the cyclotron or plasma frequency. It is proportional only to the magnetic flux density.

On average, however, the random motions of the electrons will mean they enter the magnetic field at an angle. They are forced to move in a spiral pattern along the field towards one of the poles. On approaching the poles the field strength increases, forcing the cyclotron frequency to increase and its rotational diameter to decrease, by the increasing Lorenz force on the electron but at the expense of its forward velocity. The sense of direction of the Lorenz force is parallel with the magnetic field and acts towards the region of least field strength. The forward velocity eventually reaches zero at a point known as the mirror point, when reflection occurs, and the electron reverses direction and proceeds the other way until encountering the opposite pole, where it is reflected back again. The result of this is why planetary radiation belts are doughnut-shaped toroidal regions centered on the magnetic equator.

Since circular motion created by a force involves acceleration, the electrons emit radio energy. The frequency of the emission for these relatively slow non-relativistic particles is close to or equal to the cyclotron rotation frequency. For fast relativistic particles, the radio emission also occurs at higher harmonics of the cyclotron frequency, and we call this synchrotron radiation. The radiation pattern is beamed in a thin-walled conical distribution in the same direction as the instantaneous velocity, tangential to the curve of motion (Fig. 2.3).

Radio emission from cyclotron and synchrotron sources is polarized; however, the type and sense of polarization depends on the observer's line of sight. Equatorial emission is linearly polarized where the line of sight is parallel to the electron circular motion, while near the poles the emission is more nearly circular, either left or right hand, depending on the direction of movement of the electron along the field lines. For most emissions, the viewing angle is somewhere in between these extremes, producing left- or right-handed elliptical polarization.

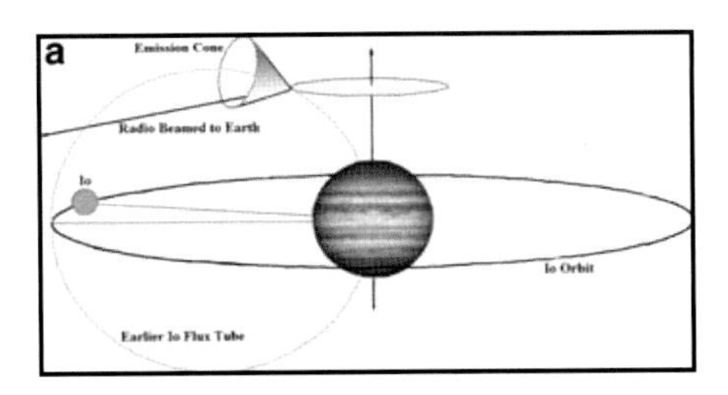

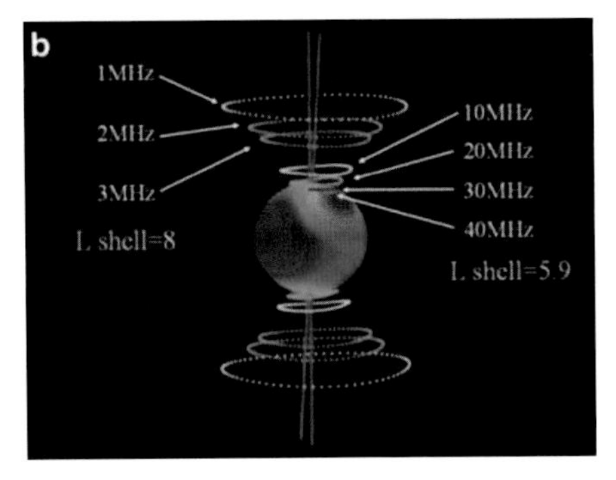

Fig. 2.3. (**a**) Drawing of the decametric radio beam emission from Jupiter. (**b**) This shows the spatial distribution of the radio spectral output. (Image credit Imai Lab, Kochi National College of Technology).

The inner magnetosphere is the source of all the strong decametric (tens of meters wavelength) radio emission, with an interesting relationship to the moon Io. Since the magnetic field is rotating with Jupiter, the field sweeps past and overtakes Io; in so doing a voltage of 500,000 V is induced between Io's inner and outer faces. In turn this leads to currents flowing in a magnetic flux tube, which connects Io to the polar ionosphere of Jupiter. The estimated current amounts to 2.8 million amperes, with a power of approximately 10^{12} W. For reasons still uncertain, plasma instabilities can occur, resulting in the strong outbursts at frequencies up to a cut off at 40 MHz. These outbursts are modulated somewhat by the motion of Io, and its mutual location with respect to Jupiter and Earth. This modulation is not completely predictable, although it helps to provide a probability of being able to detect decametric radio bursts for a given date and time.

The 40 MHz cut off is quite sharp. This is because the electrons involved in the emission process encounter their strongest magnetic field near the mirror points, after which they reflect back into weaker magnetic fields again.

The Middle Magnetosphere

The middle magnetosphere extends from $6R_j$ to in around $50R_j$. It is bound on the inside edge by the region where the influence of the planetary magnetic field is no longer dominant, and on the outside by the region where the influence of the magnetopause and solar wind is small. The dominant feature driving the field is an equatorial current sheet called the magnetodisc. Although the outer boundary of this region is largely symmetrical, the magnetodisc is thickened in the sunward direction.

The Outer Magnetosphere

This zone is significantly compressed on the sunward side due to the influence of solar wind pressure. The boundary between the interplanetary environment and the magnetosphere is known as the magnetopause. Once again the field is driven by currents flowing in the plasma within.

The thickness of the dayside outer magnetosphere is very variable, depending upon solar activity. There is also a huge difference in shape between the day and night sides. There exists a huge antisolar magnetotail. It is not known how long the tail actually is, as no direct measurements have yet been possible, but it is believed the tail reaches as far the orbit of Saturn. If we could see the extent of the magnetosphere with our eyes, it would appear a few degrees across as viewed from Earth's surface. It would certainly be an impressive sight.

Jovian Radio Emissions

Radio emission from Jupiter has been detected from as low as 10 kHz and as high as 300 GHz. The spectrum can be broken up into three parts, which relate to three distinct mechanisms involved.

The high end of the radio spectrum is dominated by thermal blackbody radiation, while synchrotron output from trapped high-energy relativistic particles accounts for the spectral region from 4 GHz down to around 40 MHz. The cyclotron process is responsible for the low-frequency output. This is one of few objects in the universe where cyclotron emission is easily observable to us.

Often in the literature acronyms such as KOM, HOM, DAM, and DIM are used, referring to the wavelengths of the radio output as detailed in the Table 2.2.

Kilometric and hectometric emissions are unable to penetrate the ionosphere of Earth, as they are reflected away and can only be studied using spacecraft. Much of this radiation is highly directional and is beamed from the polar regions, in such a way that Earth does not even intercept it. Study at these wavelengths has been carried out by spacecraft such as *Pioneer 10* and *11*, *Voyager 1* and *2*, *Galileo*, *Ulysses*, and *Cassini*, in close proximity to Jupiter

Decametric wavelengths below about 20 m and decimetric radiation can easily be observed from Earth's surface. Decametric studies are well within the capabilities of basic amateur-built equipment.

Although Earth's ionosphere always reflects long wavelength radiation, the ionospheric cut off is somewhat variable, depending on solar activity and time of day. At times of high solar activity the cutoff point is at shorter wavelengths near 10 m, and when quiet or at night the cutoff point is at much longer wavelength, around 30 m. Jovian decametric noise is known to peak around 8–10 MHz (roughly 30 m)

Table 2.2. Nomenclature of radio emissions based on their wavelength

Kilometric (KOM)	Thousands of meters wavelength
Hectometric (HOM)	Hundreds of meters wavelength
Decametric (DAM)	Tens of meters wavelength
Decimetric (DIM)	Tenths of a meter wavelength (cm)

but is still potentially very strong in the 12.5–16.5 m (18–24 MHz) band that is more accessible to a radio telescope.

Decametric Radio Bursts

When observing with a single radio telescope, due to the poor resolution compared with optical telescopes, it is not possible to resolve fine details on the surface of Jupiter. The beam width of a single HF antenna is likely to be at least several degrees wide. Even using interferometers would not provide resolutions good enough without huge baselines and aperture synthesis techniques.

In order to try and determine more information about the source location of decametric emission, shortly after their discovery, a neat trick was used. By plotting the occurrence of decametric emission against central meridian longitude (CML) over a period of a few years, an interesting profile appeared. The resulting graph, Fig. 2.4, provides a probability of being able to observe decametric emission for a given CML. The various peaks were given identification labels A, B, and C, and with later studies D. Although this technique helps to get around the poor resolution of a single receiver, it provides no information on the latitude of the source.

Although the pattern observed is repeatable year after year, even the principal peak A only occurs in about one out of three of the expected times.

In addition to the CML, it was also noticed that the phase angle (the angular separation from CML) of the moon Io had an influence on the probability of receiving HF radiation. By re-plotting the occurrence of decametric radiation with Io phase against CML the results are again interesting; see Fig. 2.5.

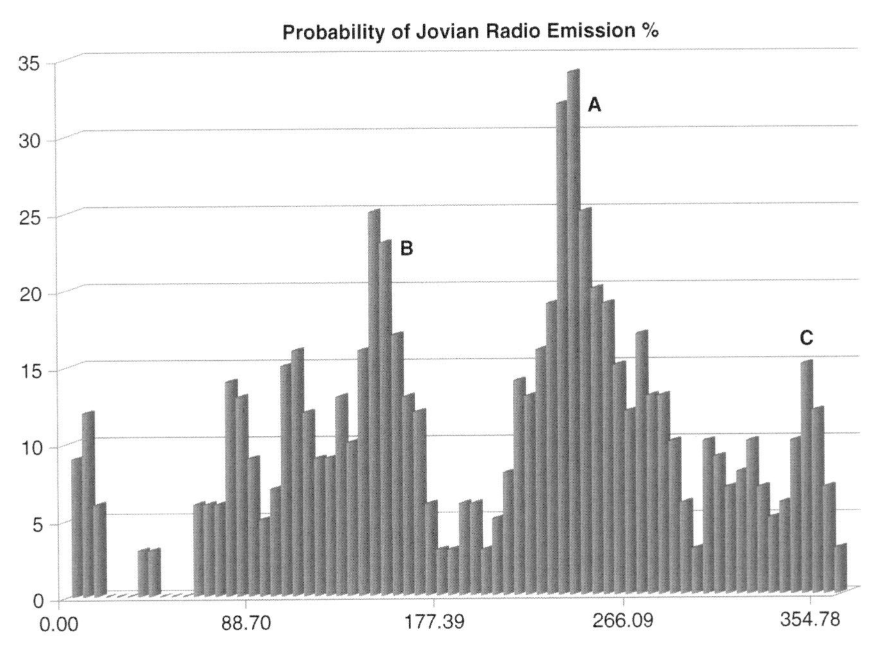

Fig. 2.4. Probability of the occurrence of decametric radio emission from Jupiter plotted against Jovian System III longitude.

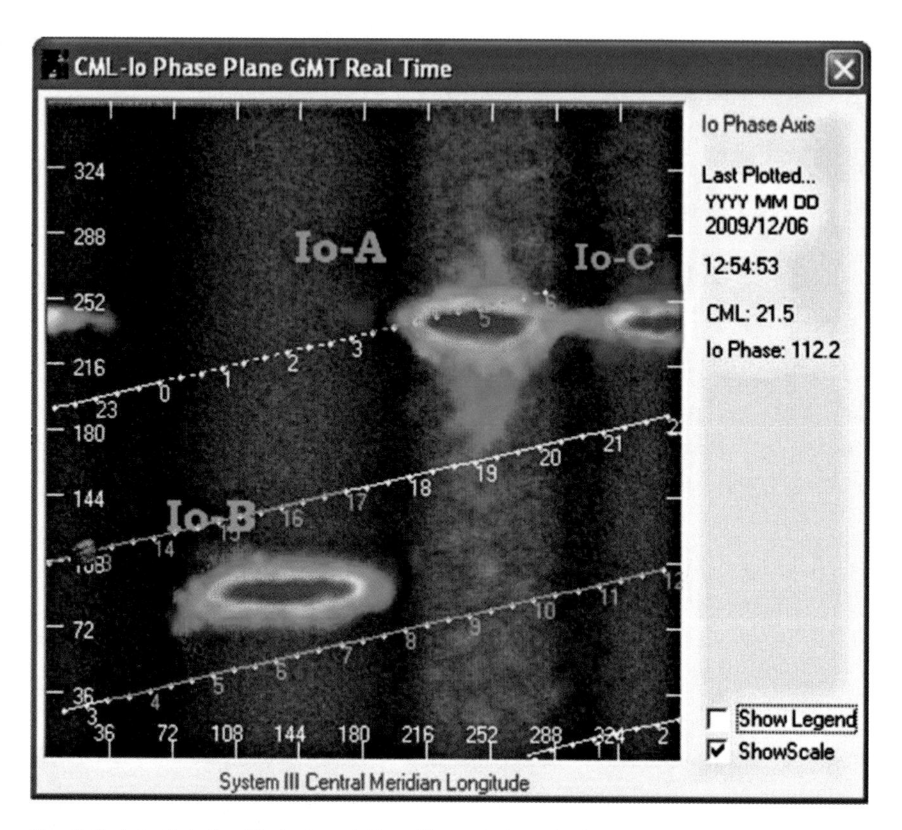

Fig. 2.5. A plot of Jovian Central Meridian Longitude (System III) against Io phase angle. This image was captured from the software package Radio Jupiter Pro-3 published by Jim Sky of Radio Sky Publishing (http://www.radiosky.com). It shows the probability of receiving Jovian DAM for your observing site at a given time and date, a highly recommended piece of software for the regular radio observer. The *red areas* indicate the highest probability.

From this diagram the oval regions are referred to as Io sources, such as Io-A source. However, decametric emission can occur at any time, particularly when the Jovian CML is in the same range as the Io-A and Io-B sources. This emission is known as the non Io-A, and non Io-B source (Fig. 2.6).

The *Galileo* spacecraft showed there is also a much weaker modulation created by Callisto and to a lesser extent by Ganymede. Further evidence of interaction by other moons come from ultraviolet images from the Hubble Space Telescope and from infrared images from the NASA Infrared Telescope Facility on Mauna Kea. In these pictures, hot spots can be seen in the Jovian ionosphere at the footprint of the flux tubes emanating from the moons, including Europa.

Jovian decametric radio emission occurs in "storms" that can easily rival Type III solar outbursts for Earth-based observers. They can last for a few minutes to several hours. The storms consist of three types of features: L-bursts, standing for "long," which is a bit misleading, since they last from about 0.1 s to a few seconds; S-bursts have durations of from 1 to 200 ms. The final type is known as N-band.

Although *Pioneer 10* and *11* offered the first chance to observe Jupiter from close proximity, the first detailed in situ radio observations were carried out by the

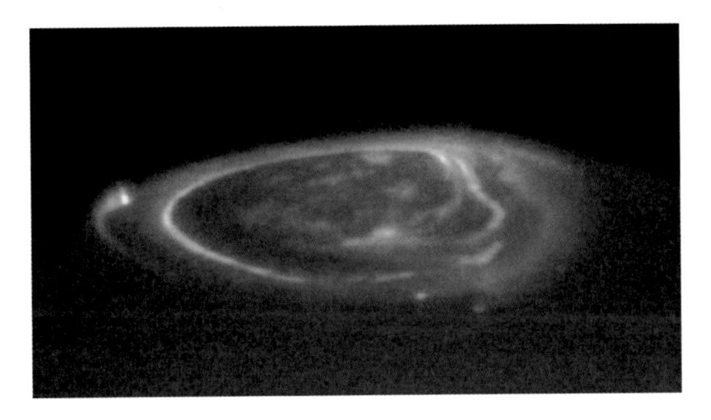

Fig. 2.6. Hubble photograph of the Jovian aurora showing the hot spot (*left side*) caused by the base of the Io flux tube, consisting of a current of charged particles in the order of one million amperes. A pair of weaker flux tubes from other Jovian moons appears in the lower foreground. (Image courtesy of NASA/ESA, John Clarke, University of Michigan).

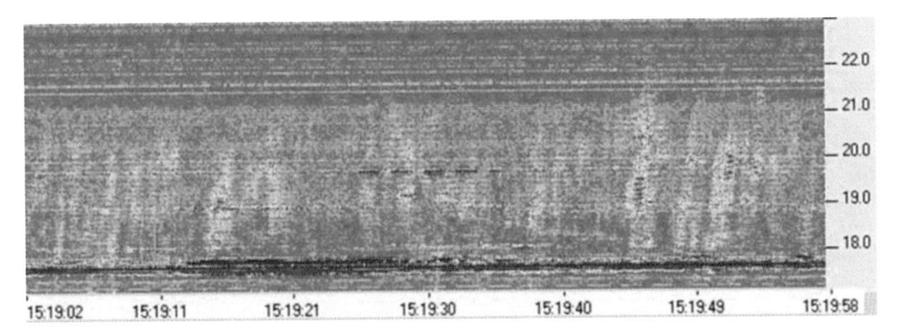

Fig. 2.7. Jovian decametric spectrum at 20.1 MHz, captured with a Radio Jove receiver by amateur astronomer Wesley Greenman on April 11, 2009. It shows L-burst activity from the Io-B source. Note the *horizontal lines* are noise or other radio activity. The *broad vertical patches* of the DAM are still clearly visible.

Voyager 1 and *2* spacecraft, which carried a plasma wave experiment for observation of low-frequency plasma waves up to 56 kHz and swept radio frequency receivers operating up to 40.5 MHz, slightly higher than the magnetic cutoff for decametric emission.

The resulting radio data was plotted on a waterfall spectrum, a type of spectral plot where one axis is frequency and the other is time. The density of the plotted results equates to relative signal strength. A modern example of a Jovian decametric spectrum taken by amateur astronomer Wesley Greenman is shown in Fig. 2.7.

The *Voyager* results showed what we now call spectral arcs – curved structures beginning with vertex early and followed later by vertex late arcs in the waterfall. The same structure can be observed by ground-based radio telescopes, but it is much more of a challenge due to the scintillating effects of the interplanetary medium (IPM), and Earth's ionosphere. The IPM and the ionosphere of Earth act on radio signals in a similar way to the way Earth's atmosphere acts on the optical image, creating a blurring effect. In addition, intervening magnetic fields, including Earth's, create Faraday rotation, which influences the polarization of the source radiation. The effect of this act is to disturb much of the spectral structure we would otherwise be able to see from an Earth-bound perspective.

Interpreting the Jovian Decametric Spectrum

The primary mechanism for decametric radio output is coherent maser emission by excited electrons. The radio frequency is close to the cyclotron rotation period of the electrons, as they revolve in spiral paths around magnetic field lines. This implies the particle velocities are much less than that of light. The emission is beamed perpendicularly to the magnetic field in a thin-walled hollow cone whose half angle is 70–80° and somewhat dependent on the radio frequency. Estimates obtained by simultaneous observation with *Cassini* and the WAVES spacecraft showed the thickness of the cone wall to be 1.5 ± 0.5°.

The Io sources are generated by the motion of Io through the inner magnetic field, which in turn induces instabilities in the plasma torus, generating Alfvén waves that propagate away from Io at an angle to the magnetic field. The exact angle varies with the local plasma conditions. Alfvén waves are a form of plasma wave. They are created when the local plasma is compressed, for example, by the motion of Io through the Io torus. The compression of the plasma induces electrical forces between the particles, which oppose and push back the plasma, which then overshoots their equilibrium position. The result is a back and forth longitudinal oscillation in the charged particles that propagates outward, similar to sound waves. Ultimately the longitudinal plasma wave can be converted into transverse radio emission by its encounter with large-scale inhomogeneities or more likely by scattering processes from small-scale local clumps of matter. The radio frequency emitted by scattering from ion concentrations is the same as the plasma frequency. If, however, the scattering is from a clump of electrons, the emitted frequency is double the plasma frequency. The conversion process to radio energy is very inefficient, however.

Theories proposed by Gurnett and Goertz suggest that the Alfvén waves undergo multiple reflections from magnetic pole regions, possibly as many as 9 times, creating a standing wave moving with Io. On each reflection the waves induce bursts of radio emission from high-latitude regions near the north and south poles. The bursts each generate an arc structure in the spectral waterfall as the cone beams sweep past the observer. The bursts are typically separated by 10° of longitude, or approximately 35 min in time in agreement with observation.

In a counter theory proposed by Smith and Wright, they believe the Alfvén wave speed would hugely increase while passing through the Io torus, thereby creating significant reflections. This would mean only about 30% of the waves would actually reach the poles. They proposed that the motion of Io creates a large magnetic disturbance in the wake of the moon, exhibiting large-scale structures around 60° wide and fine scale structures in the order of 6°. Their theory also predicts burst emissions that match observation.

Evidence that the burst emissions come from high latitudes include the greater degree of polarization for Io sources above 15 MHz. (The polarization below this level is more random.) Io-A and Io-B sources provide a significant level of right hand polarization, which is consistent with a northern hemisphere origin. Io-C and D sources give a left hand polarization, which is consistent with a southern hemisphere origin. Further studies by Queinnec and Zarka using data from the WIND spacecraft and the Nancay decametric array telescope showed that Io-B and D occur when Io is near to the Jovian dawn limb, while Io-A and C occurred near the Jovian dusk limb.

The non-Io sources have no connection with the phase of Io, or any other moon for that matter. Evidence obtained from the *Cassini* and *Galileo* spacecraft suggests that the non-Io sources occur at extremely high latitudes, well above the Io flux tube footprint in the ionosphere. These are believed to occur in a region of the outer magnetosphere in a zone closely tied to the magnetotail. The triggering mechanism is thought to be interplanetary shock waves propagated by the solar wind. The shock waves trigger the sudden release of magnetospheric energy, creating decametric and hectometric burst emissions.

Further evidence of the solar influence on non-Io sources comes from the correlation with the 11-year solar cycle. The occurrence of non-Io DAM varies linearly with the size of the Sun's polar coronal hole.

Chapter 3

Meteors and Meteor Streams

Meteors and meteor streams are a fascinating subject to many amateur astronomers. Their sometimes spectacular transient appearance is like nothing else in astronomy. Although meteors do not emit radio waves, their highly ionized trains are quite efficient at reflecting VHF radio waves. In this chapter we look at the nature of meteors, their motions and origins.

Meteors occur when small dust particles enter the tenuous upper atmosphere at high velocity. Friction with the air molecules creates enough heat to vaporize the meteoroid and form an ionizing column of air in its wake. There is a constant sporadic background of activity, which is enhanced at certain times of the year when Earth passes through streams of particles.

Stream-based meteors appear to radiate in all directions from a small region of the sky known as the radiant. This is simply a perspective effect, in the same way that a straight motorway viewed from the center of a bridge appears to get narrower into the distance. The radiant is not, however, fixed. It moves constantly, usually drifting approximately 1°/day as Earth moves through the stream in its orbit around the Sun.

Meteor streams are named after the constellation in which the radiant is located (at peak activity). For example the Perseids are named after the constellation of Perseus, the Taurids after the constellation of Taurus. One odd one, though, is the Quadrantids, named after the constellation Quadrans Muralis, which no longer exists and is now part of Boötes.

In some cases more than one stream radiant occurs in a constellation, in which case the Bayer letter of the nearest bright star is added to distinguish it, for example the η Aquarids, and the δ Aquarids. Since meteor streams (often referred to as meteor showers) are the remains of comet dust left behind in the wake of a comet passing through the Solar System, a few showers are named after their parent comet. One example is the Giacobinids, named after the periodic comet 21P/Giacobini-Zinner.

In view of their cometary origin, streams can be aged to a certain degree by their rate profile and their duration. Recently deposited dust follows the comet orbit. The individual orbits of the particles are tightly packed close to that of the parent object. As Earth passes though a young stream the rate of activity builds quickly to a peak and sharply falls off afterwards. The duration may be measured in hours, or

J. Lashley, *The Radio Sky and How to Observe It*, Astronomers' Observing Guides,
DOI 10.1007/978-1-4419-0883-4_3, © Springer Science+Business Media, LLC 2010

a few days. For older streams, non-gravitational forces such as interaction with the solar wind act to spread the orbits of the particles. These streams are much wider, and the resulting meteoric activity lasts for many days or even weeks. The build up to peak activity is slower and the tail is often afterwards extended. The Perseid stream falls into this category, where detectable activity begins in mid-July, but peak activity does not occur until around August 12.

We should point out here that the Geminid stream has been associated with the object 3200 Phaethon, which was classified as an asteroid. Phaethon is considered to be the dormant core of a comet that has used up all its volatile compounds. The association of meteor streams with parent objects is mainly based on the similarity of the particle orbits with that of the parent object. It is of course possible that Phaethon is just coincidentally in a similar orbit to now-dead comet that spawned the Geminids. The only way to be sure would be to compare the composition of Phaethon by either direct sampling or careful spectral analysis using the spectrum of Geminid meteors. The emission lines seen in meteor spectra clearly demonstrate the chemical composition of the objects.

Meteor streams are quantified by a figure known as the zenithal hourly rate, or ZHR. ZHR is a theoretical calculated value of the number of meteors that would be seen by a single observer on a perfectly clear and dark night if the radiant were placed at the zenith. It is used as a way of directly comparing the hourly rates of one meteor shower with another. Observational effects must therefore be removed from the observed rates. The observational effects to take into account are altitude of the radiant, sky transparency, sky brightness, cloud cover, and obstructions in the field of view.

The calculation of ZHR from observed rates is given by:

$$ZHR = \left(\frac{N}{t}\right) RLC$$

In this equation:

- N is the observed number of meteors
- t is the duration of the observing run in hours
- R is the correction factor for the altitude of the radiant
- L is the correction factor for limiting magnitude
- C is the correction factor for cloud cover or obstructions

Determination of R

Due to atmospheric extinction, which increases with decreasing altitude, the number of meteors observable per hour increases as the radiant rises in the sky, even if the actual rate of the stream does not change. In most cases a given radiant does not pass the observer's zenith, but an estimate can be made. The correction factor is

$$R = 1 / \sin(\phi)$$

where $\phi = \textit{the altitude of the stream radiant in degrees}$

Due to the constant diurnal motion of the radiant, the count of meteors should be logged separately for each hour. The radiant altitude is then determined for the center of that hour.

Determination of L

For each hour of observation, the limiting magnitude should be determined in case the conditions slowly change. This reading should be taken at least every half an hour, and an average value used in the correction for that hour's rate. L is calculated from the following formula:

$$L = r^{6.5-LM}$$

where r is the population index for the meteor stream.

The population index for a stream is calculated from the distribution of observed magnitudes; a high value of r suggests a greater population of bright meteors. See the tables in the meteor calendar section. Older meteor streams become depleted of fainter meteors, leading to the higher r values. Younger streams have more of a balance of population across the visible magnitude range.

Determination of C

Where significant cloud cover exists, and only small regions of sky are viewable, rate counting is particularly inaccurate and is not feasible. Although an observing session may start out well, there may be periods when some cloud drifts over. An estimate of the fraction of sky obscured by cloud cover should be recorded for each hour observed. Then the correction C is calculated from

$$C = \frac{1}{1-x}$$

where x = the fraction of sky obscured. If $x > 0.2$ the ZHR will be unreliable.

Radio Scattering off Meteor Trails

As soon as the Second World War ended, scientists returning to civilian duties began using the equipment designed and built for the war effort. This included the adaptation and use of surplus radar equipment to study the properties of meteors. By transmitting a VHF signal, and observing the backscattering from meteor trains, it was possible to determine the velocity of meteors as they sped through the atmosphere. By then combining the results of a pair of radar units it was possible to determine their true path through the atmosphere, their velocity, and therefore their orbits.

A licensed amateur radio person could attempt to use radar in the same way, but the use of legal levels of output power would severally restrict the useful range, and the effort is unlikely to be very successful. However, the amateur can use forward scatter techniques to study meteors. Forward scatter involves the observer receiving reflections from meteors, where the transmitter is too far away for direct reception and is only heard when it is reflected from a meteor trail.

Although meteor scatter is possible in the HF spectrum, the ionosphere is also reflective there, too, which is counterproductive. The HF spectrum is particularly noisy, too. Best results are obtained from 30 to 100 MHz, although amateur radio enthusiasts regularly encounter activity at up to 145 MHz with reduced efficiency.

Traditionally Band I television transmitters were used by amateur observers, in the range from 51 to 83 MHz, but sadly, with the approach of the digital TV switchover, these transmitters are disappearing in favor of UHF channels. At the time of writing there are still some operating in Europe and other parts of the world. It is still possible to use the broadcast band FM radio frequencies in the range 83–108 MHz, but lower frequency is still preferable.

Specular Reflection

Radio scatter off meteor trails is entirely specular, which means that the waves follow the same rules as light reflecting from a mirror. The same geometric rules apply, too, so the angle of incidence for incoming radio waves is equal to the angle of reflection as it leaves (see Fig. 3.1).

To understand what that means in practice for the observer, consider initially the case for radar observation. To receive an echo, the wave is transmitted from and received back at the same location, which has to be perpendicular to the path of the meteor (angle of incidence and reflection of 90°). The geometry of this situation is shown in Fig. 3.2.

The geometrical orientation for the radar observer can be imagined as a series of spherical shells, where the observer is at the focal point of the shell. The figure demonstrates that all members of a given meteor stream, whose paths by definition are parallel, generate echoes in a single plane, known as the echo plane.

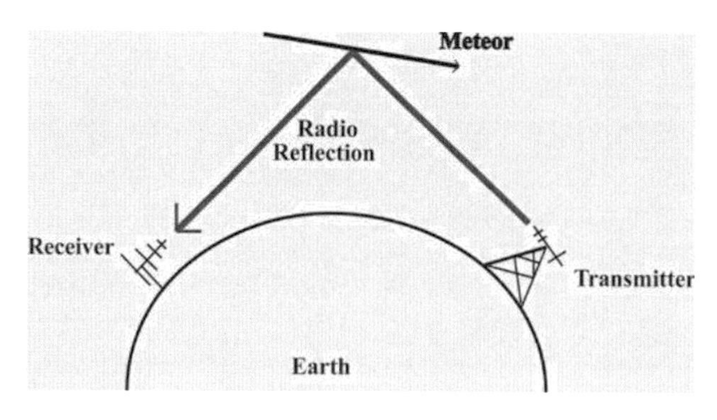

Fig. 3.1. Forward scattering of radio waves from meteor trails.

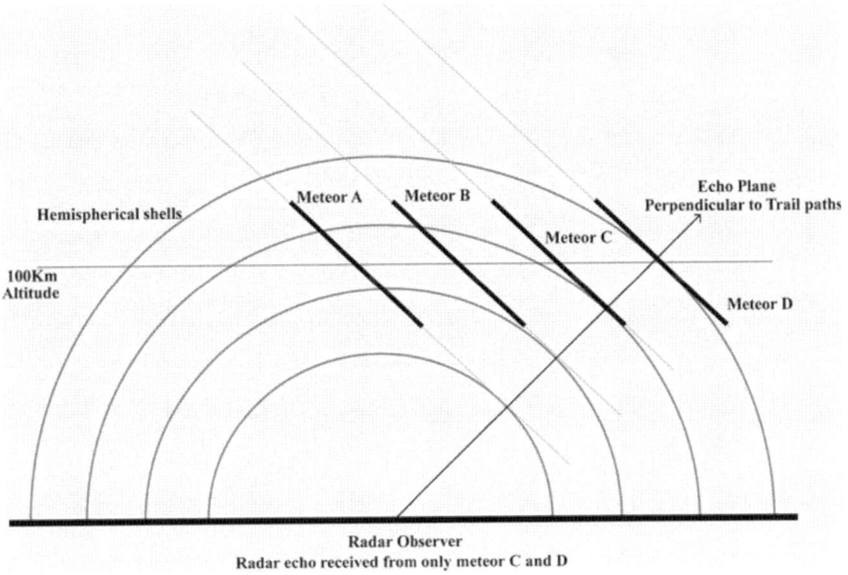

Fig. 3.2. In the case of a radar observer, only meteors occurring across the echo plane will be detected. Meteors A and B will not be seen. The echo plane is a great circle.

This echo plane projects a great circle onto the sky, so that the antenna could be aimed at any point along this great circle to receive data from meteor stream members.

The geometry for the forward scatter situation changes, however, to an ellipse where the transmitter is at one focus of the ellipse and the receiver is at the other focus (see Fig. 3.3).

If the observational baseline between the stations is fairly short, the distribution of reflection events will be fairly evenly spread along the ellipse. However, if the ellipse is long, there will be concentrations of observable echo activity nearer to the ends of the ellipse at locations both near to the transmitting focal point and to the receiving focal point. So by aiming an antenna at a quite high altitude at the receiving station these end point reflections will be obtained. However, this is not the best orientation.

Assume for a moment the sky is filled uniformly with meteor radiants, not such a bad assumption when observing sporadic background meteor activity. Reflections with the greatest signal power and duration will occur for meteors appearing over the midpoint between the transmitting and receiving stations. This is because the longer the baseline distance is between stations the greater will be the reflected power from a meteor occurring over the midpoint. This should not be too much of a surprise.

For example, take a clear pane of glass; when held perpendicular to your line of sight it is transparent. Now tilt the glass slowly. You can see some reflection from the surface. When the angle reaches a certain critical value it reflects all of the light.

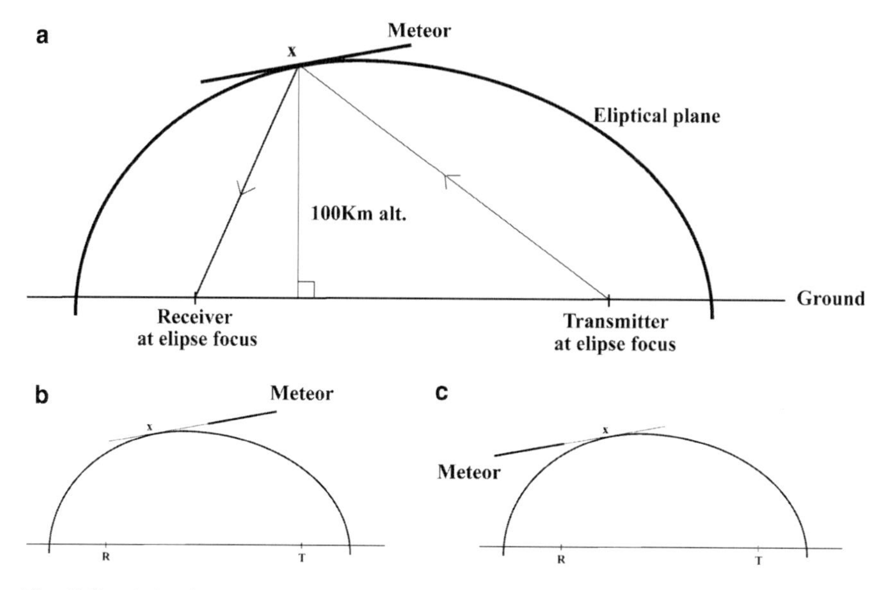

Fig. 3.3. In the forward scatter case, the echo plane is no longer (**a**) great circle. It is a band across the sky. The observer and transmitter are located at the focus of an elliptical shell, which passes through the meteor zone at an altitude of about 100 km. Only meteors occurring tangential to point x can be seen. The meteors in case (**b**) and (**c**) are not seen.

Table 3.1. Suitable antenna azimuth angles for some annual meteor streams (assuming a northern hemisphere location)

Meteor stream	Dates	Azimuth for antenna	Alternate azimuth for antenna
Quadrantids	Jan 1–4	N – NW	S – SE
η Aquarids	May 1–6	NE – E	SW – W
Arietids (daytime stream)	June 1–15	NE – E	SW – W
δ Aquarids	July 26–31	E – SE	SW – W
Perseids	Aug 10–14	NW – E	SE – W
Orionids	Oct 18–23	NE – SE	SW – NE
Leonids	Nov 14–18	NE – E	SW – W
Geminids	Dec 10–14	NE – E	SW – W
Ursids	Dec 21–23	NE – E	SW –W

The maximum amount of energy is then being reflected. This angle is relatively shallow, so in the sense of meteor forward scatter reflections this equates to a transmitting station far in the distance. However, taking into account signal attenuation due to traveling distance, there is a practical compromise on baseline distance. These midpoint reflections occur in two hot spots, roughly 50–100 km either side of the baseline center point. The echoes will be of stronger and longer duration than the end point reflections.

For example, a 1,200 km baseline would provide hot spot echo durations six times longer than a 600 km baseline. The durations over the 1,200 km baseline would be roughly 92 times longer than end point reflections.

In order to set up the equipment to efficiently study a specific meteor stream activity by radio, refer to Tables 3.1 and 3.2 The first table suggests suitable azimuth ranges for the major meteor streams. A suitable transmitter should then be

Table 3.2. Antenna altitudes and offsets for variations in transmitter range

Range to transmitter	Altitude for antenna	Azimuth offset (see text)
500	18	21
550	17	20
600	15	18
650	14	17
700	13	16
750	12	15
800	11	15
850	10	14
900	9	14
950	9	13
1,000–1,100	8	12
1,100–1,200	7	12
1,200–1,300	6	11
1,300–1,400	5	11
1,400–1,500	4	10
1,500–2,000	1	10

determined for the chosen direction. This should not be difficult for broadcast FM radio band, but could prove tricky for Band I TV signals..

Once the station has been chosen determine its azimuth bearing and range. To make this easier there are a number of websites that allow you to type in two locations and provide you with a bearing. Searching with your favorite search engine should easily find one. Bookmark it in your favorites for future use. You may even have a GPS tool that can do this for you. Table 3.2 then gives you the altitude for the antenna, and an azimuth offset from the bearing determined earlier. If the radiant has a positive declination, the offset is to the south, and if the radiant has a negative declination, the offset is towards the north.

This discussion only refers to using directional antennae such as Yagis, log periodic arrays, etc. Some observers successfully use omnidirectional antennae, which are vertically mounted and no special aiming is needed. A worthwhile experiment might be to set up two receivers, one with an omnidirectional antenna and one with a good directional Yagi antenna, and compare the results. It may take a few years of data gathering to determine whether activity of weaker showers is more noticeable in the results of the directional system.

Modeling of Meteor Radio Scatter

A mathematical model is a means by which scientists can learn more about physical processes occurring in nature. The mathematical model is generated based on observational evidence and theoretical ideas and is used to make predictions. The predictions are then compared with observations to see how closely they match reality. Close matches suggest at least the theory is correct; if not, the model is changed, and the cycle begins again.

The first attempts at modeling the specular reflection of meteor streams is the simplest to describe and works reasonably well for the extreme cases of underdense and overdense meteor trails.

Underdense Trails

As the name suggests underdense trails occur for smaller, low-energy meteor trails. The electron density responsible for the specular reflections is relatively low. This model makes several broad assumptions, as follows:

- The trail is a stationary, linear column of free electrons.
- Its diameter is small compared with the wavelength of scattered radio radiation.
- The column does not expand.
- The electrons do not recombine or diffuse.
- The trail is infinitely long.

Most of these assumptions are not true, of course. High-atmosphere winds will distort the trail. The trail will expand as it twists, electrons will recombine, and it can't be infinitely long. Only the second assumption is a good one.

Despite these drawbacks the model does work reasonably well for the extreme cases. The assumptions made are reasonable for the first few fractions of a second involved in radio specular reflection.

In the underdense case the model shows radio waves penetrate into the electron column and excite the electrons into physical oscillation. Electron collisions with other particles are assumed to be zero, so they do not recombine. Whenever an electron is subject to an accelerating force it will radiate radio energy; therefore electrons re-radiate in all directions. Secondary effects are also ignored, so that each electron acts as if there are no others around, although radio is scattered from all parts of the trail and electron-generated energy is emitted in all directions. Constructive interference only occurs in the direction predicted by geometric optics (for reflection off mirrors).

In fact, it turns out that the simple model is not that bad. It can explain the properties of the low energy underdense trails and the very high density overdense trails, but it breaks down for the intermediate cases.

Radar echoes from underdense trails have a fast rise time, a short plateau, and an exponential decay. If the radar system is coherent (so that the transmitted pulses are always in phase relative to each other), the reflections maintain the phase of the pulses. The plateau in the received echo will then exhibit oscillations (caused by interference) on a millisecond timescale, which contains useful information about the velocity of the meteor.

The model only works for a short length of trail centered on the closest point of contact to the receiver. This is not a problem, because the extreme ends of the trail contribute little to the received echo. The model successfully demonstrated that the received echo power will increase with increasing wavelength (decreasing frequency).

The Overdense Trail

For high electron densities, the central core of the trail will be plasma or close to it. In a plasma the electrons are entirely stripped off the atoms. In this case radio energy can't penetrate into the trail and is scattered from the surface. The classical model assumes the electron density is uniform across the trail, but in practice it has a Gaussian distribution across the diameter.

The above assumption does not take into account penetration of radio into the lower density outer regions, or the effect of winds distorting the trail.

In the most extreme cases, the radius of the trail is significant, which means the amount of reflected power for a radar system is directly proportional to the radius of the cylinder. However, this breaks down for lower energy overdense trails, where the radius is small. This model is less useful in the forward scatter case, where extreme reflection angles are often involved.

The definition of an overdense meteor trail is when the electron density exceeds 2.4×10^{14} electrons per meter length of trail. The classical theory works well except for the transition region between the underdense and overdense case. The full wave theory gives overall better results but is much more complex.

Forward Scatter Radio Reflections

The previous discussions on underdense and overdense meteor trails were based on theory developed for backscatter radar observation methods. In this section let us consider the radio reflection by forward scatter. As amateur astronomers this is our only means of studying meteor activity by radio.

Figure 3.4 shows two reflected signals from a meteor trail. Trail y has a longer distance to travel than trail x, but they were both emitted from the same source at the same time, so they started out in phase. However, the different path lengths

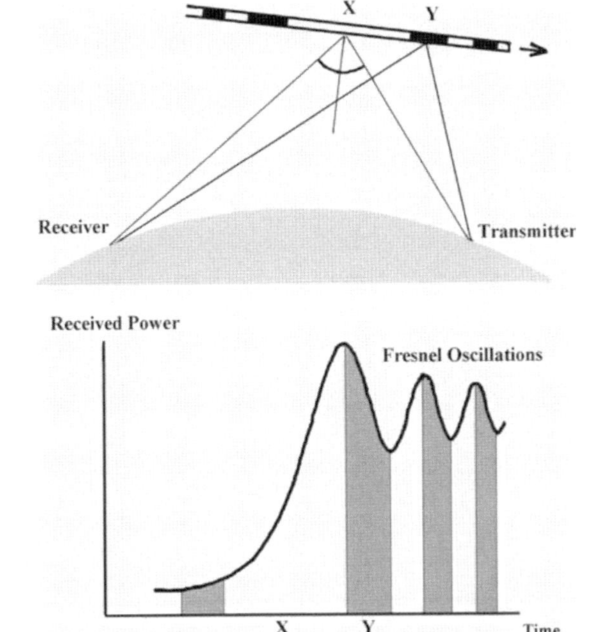

Fig. 3.4. Fresnel oscillations. When multiple reflections are received from different parts of the same meteor trail, interference occurs. When the total path length (in units of radio wavelength) of x and y are similar, reinforcement occurs, and when they are different minima occur.

from the different reflection points mean the radio energy can arrive in phase or out of phase. The receiver therefore sees an interference pattern.

A radio echo first quickly rises to its peak power level. Then as it develops it oscillates and decays. The regions of meteor trails producing in phase and out of phase reflections are known as Fresnel zones. The spacing of the Fresnel zones, and therefore the associated oscillations, carry information about the velocity of the meteor. As in the case of backscatter systems, the measurement of the oscillations requires the data capture system to have a sampling rate of at least 1,000 times/s. The full geometry of the reflection path must also be known to calculate the meteor velocity.

In practice the determination of the oscillation frequency is made much more difficult due to several factors. Distortion or break up of the trail by winds alters the angles of reflection and the resulting interference pattern (mainly affecting long-duration overdense trails). Also deceleration of the meteoroid and the diffusion of the ionized particles complicate matters.

Diffusion of the particles in the trails of meteors will be greater for higher altitudes because the lower density of the atmosphere allows for a greater mean free path for particles. The initial radius of the ionized trail will therefore be greater at higher attitudes than lower down. Taking the case of an underdense high altitude meteor, the initial radius of the trail may be large in comparison to the wavelength of radio frequency used to observe it, thereby seriously reducing or even preventing reflection. This means there is a maximum altitude beyond which it is impossible to detect underdense meteors, known as the echo ceiling. Small, fast meteors that begin their ablation at high altitudes are discriminated against in radio observation techniques. Overdense trails are not affected, because the reflection is from the surface of the trail, and not from within the body of it.

Forward scatter methods have an advantage over radar techniques in that reflection is detectable from underdense trails with a larger radius, and therefore higher altitude.

For a more in depth coverage of the physics of meteor trails and radio scattering refer to the book *Meteor Science and Engineering* by D.W.R. McKinley, which is still one of the best sources available despite its publication date of 1961.

Setting Up a Meteor Scatter Radio Receiving Station

Choosing a Receiver

The radio should be capable of receiving VHF channels from 30 to 108 MHz or even higher. Ideally, it should be a multimode receiver, in other words, capable of receiving signals modulated in the following ways: amplitude, frequency, single sideband (upper and lower), and CW (or Morse). It is important the receiver be sensitive, selective, and stable. For more information on these subjects refer to the radio theory chapter of this book, where receiver specifications are discussed in detail. Model-specific advice may be obtained from local amateur radio clubs; by getting involved in these groups you may be able to try before you buy.

As an alternative to obtaining a wide band receiver, a good quality traditional communications receiver – often called short wave receiver (which covers frequencies up to 30 MHz) – may be used with the addition of a frequency converter to allow it to be used for VHF work. Ideas for constructing frequency converters are covered in the later project chapters of this book. This author, for example, has an Icom IC707 HF amateur radio transceiver. Although this can only transmit on legal amateur bands, it can receive continuously up to 30 MHz. It made sense to construct a frequency converter for meteor scatter experiments. The receiver is stable and multimode capable. So long as the frequency converter is also stable it makes a useful platform for many experiments.

Although there are many receivers on the market capable of receiving HF, VHF, and UHF bands, not all offer multimode options. Some of the receivers (often referred to simply as scanners) don't offer all the modes in all the bands; for example, single side band mode may not be available for the commercial FM broadcast frequencies but present on the HF bands. Clearly there is logic there. No one transmits side band channels in the 88–108 MHz region!

A useful receiver for many radio projects would be the Icom PCR1000. It is a computer-controlled radio covering a wide range from low kHz up to 1,300 MHz and fully multimode. It is not restricted in the VHF bands – any mode of reception can be selected. Many meteor scatter observers use these radios. Sadly it is no longer manufactured. The most important feature for radio work, the AGC (Automatic Gain Control) on/off feature, is not present in its successors, the PCR1500 and PCR2500. The later models list the AGC function as fast or slow, but not off. It should still be possible to easily source a PCR1000 from second-hand dealers, on line auctions, or private sales through amateur radio clubs or magazines.

The ability to switch off automatic gain control is very important in radio astronomy. AGC alters the signal amplification based on signal strength. We need to monitor the raw signal without any compounding instrumental effects (Fig. 3.5).

In order to gain some experience before committing to buy a good multimode receiver, you can start with a good-quality broadcast band FM receiver. You may already have one. While testing such receivers for this book, I needed to make a quick modification to the receiver so I could adequately connect it to my antenna. The Antenna available was a log periodic array, a type of broadband but directional antenna. The fitting was N type, as the antenna was suitable for use up to 1,300 MHz, where the quality of the N type connector was needed. However, the receiver had the standard coax socket of the type only used by commercial FM receivers and hi-fi systems. Since the receiver was obtained free of charge from a friend and was not used for regular listening, I was happy to modify the unit for its new found use. The coax socket was removed and replaced with a chassis mounted BNC socket. A suitable length of coaxial cable was at hand, to which was fitted a BNC plug at one end and an N plug at the other. I always carry some spare coax and a range of plugs for just such eventualities. Experiments during the Perseid meteor shower yielded some success in picking up strong reflections. I took the audio output of the radio to a PC sound card and used Spectrum Lab software to observe it. The interesting thing was I got inverted traces! The average level dropped when an echo was received. This was due to the high level of background hiss you get from an FM receiver when no signal is received. This generated a fairly constant background trace on the audio spectrum. When an echo was received a

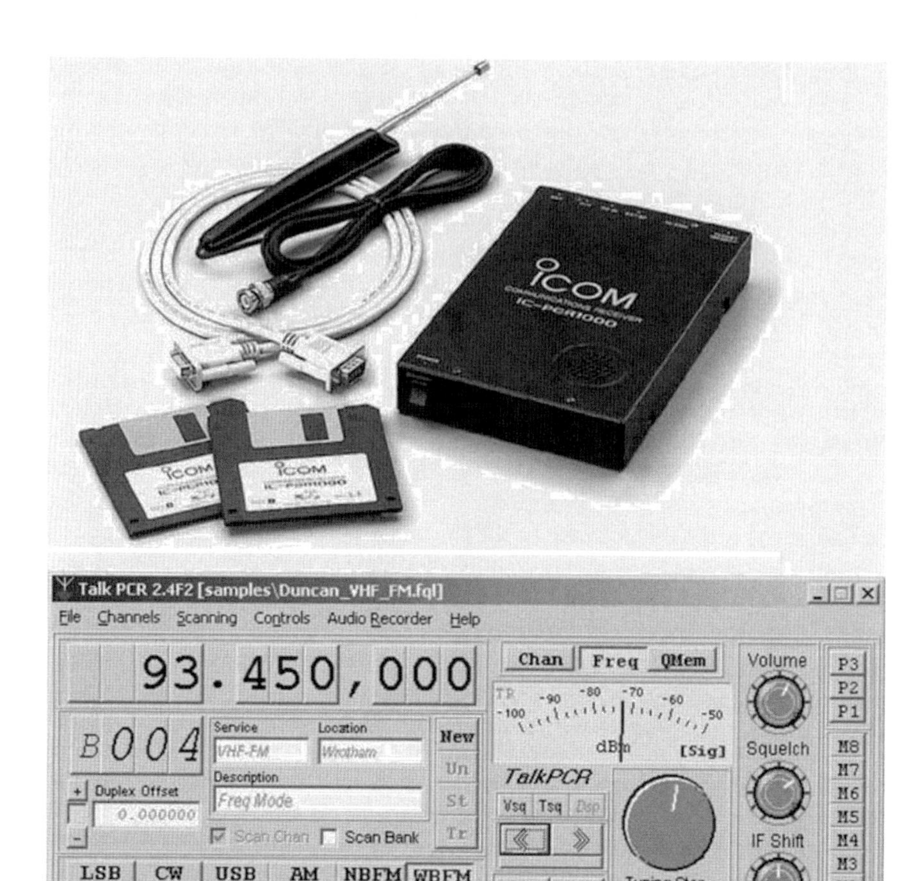

Fig. 3.5. The Icom PCR-1000 and its software control interface.

short burst of radio signal from the station was heard with a very quiet background level, so the average noise dropped. Spectrum Lab is a very useful piece of software, and its use is discussed in later chapters of this book.

The Antenna

The antenna in any radio system should be of the highest quality, or constructed with the greatest of care and attention. A good antenna constructed specially for the operating frequency and carefully impedance-matched to the receiver will most efficiently pass collected radio signals into the receiver. For more information on choosing or building antennae refer to the antenna chapter of this book. Use Yagi's if you set up a station for a fixed frequency and orientation. If, however, you plan to optimize the antenna aiming in order to best study individual meteor streams, it would be an advantage to use an antenna with a wider frequency bandwidth, such as the log periodic array.

Choosing a Radio Channel

The website http://www.ukwtv.de/tvlist, which at the time of writing was last updated in 2001, contains a listing of a large number of television transmitters around the world. Some further research, however, will be needed to determine whether a given channel is still operating on the lower bands. Another useful source of channel information is the World radio and TV handbook (WRTH), which is published annually. If you are planning on trying the FM broadcast band from 88 to 108 MHz band, the following website is useful for selecting frequencies: http://www.fmlist.org.

Observing Technique

It is important initially when establishing the meteor scatter station that you take time to listen manually for meteor activity. Only by doing this and learning what genuine echoes sound like can you then set up an automated way of counting these events. It is useful to be able to listen to and log data simultaneously, so that the echo profile of genuine events can be seen and heard. This is straightforward when using a PC sound card as the analog to digital converter interface. The PC is the only way to constantly capture and analyze data.

The reason why a multimode radio receiver is recommended for this work is because you need the single side band (this may be upper side band USB, or lower sideband LSB) mode of operation. Nearly all SSB receivers also have at least AM mode alongside, and some have FM, too. We will use the SSB mode regardless of the type of modulation employed by the transmitter, which in most cases will be frequency modulated.

The key thing to remember here is that you are not trying to receive the channel in the normal way. In fact this is not helpful. You want to study the reflection profile of meteor trails. The desired source transmission is an unmodulated carrier wave whose amplitude is constant. That way any changes in amplitude received is due to the properties of the echo. An FM radio or TV channel carrier wave has constant amplitude. An SSB demodulator is based on a form of amplitude modulation; the difference is that an SSB signal has no carrier component. The "carrier wave" is regenerated inside the receiver by what is sometimes referred to as a beat frequency oscillator. The beat frequency oscillator signal is mixed with the radio frequency, generating a sum and a difference of the two (see the radio theory chapter in this book for more information on mixers).

When the adjustable beat frequency oscillator is set close to but not quite equal to RF signal the difference result is an audio tone. The pitch of the tone depends on how close the beat oscillator is to the RF signal. (In practical receivers the demodulator does not work on the original radio frequency but on an intermediate frequency, which is usually lower. The same principle applies because the IF frequency contains the same information as the higher radio frequency.)

So the method of using an SSB receiver to monitor FM channels will provide a tone when a meteor echo occurs. This tone sounds like the "pling" sound of a tuning fork for underdense trails. And can sound like a "bong" sound for overdense trails. The intermediate case known as a transition echo can start out like an overdense dense case, but then it begins to oscillate as it fades.

Underdense trails are generated by fainter meteors, the kind that would be at the limit of visibility to the naked eye, or below. The overdense and transition events are caused by meteors that would be easily visible. However, this is not to say you would be able to see the meteor that creates the overdense trail, if you are observing the midpoint hot spots. These events will be occurring a few 100 km away from you and would not be directly visible, as they would be attenuated by atmospheric extinction close to the horizon.

On the whole the fainter meteors far outnumber the brighter events, so underdense echoes are the most common. In a few rare cases it is possible to observe the head echo of the meteor, which sounds like a descending pitch whistle caused by the Doppler shift of the signal. This is usually followed by either a strong underdense echo, or the bong of an overdense event.

The previous discussions fully rely on the SSB observing technique. Using a conventional FM receiver makes this impossible, and only rate-counting studies can be done. The brief reception of audio from the transmitter is compounded by the variability of the speech or music contained in the signal.

Automating the Observations

Clearly a full 24 h, every day of the year data capture system must be able to automatically reduce and store the data. Old-fashioned methods of data capture such as paper chart plotting requires manual interpretation. The data will accrue far quicker than it can be reduced manually!

The only feasible way to capture and analyze meteor data is by using a computer. One method involves a standard sound card as an audio analog-to-digital converter. The specification of the computer is not at all critical. Older PC's such as Pentium II, III, and IV-based systems can easily be obtained cheaply and sometimes even free of charge. These can be dedicated to one task of logging and analyzing radio astronomy data. At the time of writing this author has eleven computers, several of which are simply in store waiting for tasking.

Software to analyze the collected data could be custom written. Consider using National Instruments LabVIEW software, as well as Mathworks Matlab software and an open source language called Processing. Full discussion of these packages (free alternatives are Octave and Python using libraries such and numpy and scipy) is beyond the scope of this book, but they offer very powerful tools for data capture and signal processing. The software probably best to begin with is Spectrum Lab written by German amateur radio enthusiast Wolfgang Buescher, whose amateur call sign is DL4YHF. The software is freely available from his website: http://freenet-homepage.de/dl4yhf/spectra1.html.

Spectrum Lab is an audio spectrum analyzer and has a built-in scripting language that can be used to identify meteor echoes, count them, and even save a portion of the audio spectrum where specific conditions occur. See the chapter on data logging for details. Andy Smith is a keen active amateur radio astronomer in the UK who is constantly monitoring meteor activity. He offers assistance for this on his website: http://www.tvcomm.co.uk/radio/how-to.html.

The basic principle in setting up spectrum lab to log meteor activity is to define an amplitude threshold condition. When the signal exceeds this threshold value then it can increment a count value. Its duration can be determined and its spectrum recorded. This could pose a problem for transition meteors. The strong oscillations present in the signal may drop below your chosen threshold, and each oscillation may then be counted as a different meteor. The FM receiver technique may also give you inverted traces, because the background hiss of the FM receiver drops away as each piece of radio transmission is received.

There are times when meteor rate counting will be impossible. The phenomena known as Sporadic E and D layer scattering can create reflections for VHF signals that are not associated with meteor activity. Certainly where continuous reception for many minutes (say more than 15 min) is concerned it is most likely due to Sporadic E. This is caused by unusual ionization levels in the E layer of the ionosphere. Although it can occur almost any time of year, it is more common in the summertime during the day. For northern hemisphere observers the probability of occurrence peaks in June, reducing in July and August. During these spells VHF propagation is significantly enhanced for up to several hours at a time, and stations that would not normally be heard can appear as if they were transmitted locally.

Another possible noise source to watch out for is thunderstorms. You can't rely on hearing thunder, either, because the radio noise from thunder can propagate just as well as meteor echoes. This may cause an error in a rate-counting system. Although the audio spectrum of thunder will look very different than that of meteors, it is still hard to discriminate by profile in an automated way.

One way to avoid the confusion is to simultaneously log an empty radio frequency in the range of 2–10 MHz. Where an event occurs simultaneously at both frequencies then thunder is the most likely cause, and this should not be counted by the software. The use of the left and right channels of one sound card can allow both to be monitored together.

How to Confirm If Your System Is Working Properly

When you first set up a new system, you need to be sure it is working correctly. As mentioned earlier both listening and logging can help to confirm this, but you are not always going to be there listening. Are there significant random sources of noise? Is thunder a problem? Tropical areas suffer from regular thunder activity at certain times of day and year.

Well, meteor activity is clearly not uniform throughout the year. You will expect the peaks of activity near stream maxima, but there are other more regular variations in meteor detectability. Firstly there is a significant diurnal variation. Peak activity is expected between midnight and 6 a.m. Rates then fall throughout the day up to 6 p.m., and then they begin rising again.

The reason for diurnal variation is simple. The observer's hemisphere is facing the direction of motion of Earth in its orbit at night. The relative impact velocity is therefore enhanced, generating more friction and greater levels of ionization. The effect occurs for both visual and radio observers.

The second variation is annular, with increased rates of sporadic meteors in the 6 months from July to December. The increase is fairly small compared with the diurnal variations, about 1.31:1. Part of the reason for the annular changes is known as the apex of Earth's way. It is the projection of Earth's orbital motion against the sky and is approximately 90° ahead of the Sun. So that in the latter months of the year this appears in the observer's nighttime sky with again enhanced impact velocities. However, even after removing these effects, there is still a component of annular variation in sporadic rates that is due to the non-uniform distribution of meteoric particles in the vicinity of Earth's orbit. This is weak evidence towards the cometary origin of sporadic meteors.

The Annual Meteor Streams

The International Meteor Organization (IMO) maintains a working list of meteor streams and publishes an annual calendar of the observational prospects for that year. The aspects of observability are aimed mainly towards the optical techniques. Clearly radio astronomy is unaffected by moonlight or even daylight. The working list does not cover all known streams. The weaker ones, which are difficult to observe visually are omitted. Radio detection at low frequencies of around 30–50 MHz are particularly good for detecting underdense (as well as overdense) trails from meteors that would be missed by visual techniques. Sadly, the higher frequencies in the range of 88–108 MHz are not very good for detecting underdense events and will almost exclusively yield overdense results.

Annual Meteor Calendar

The table data included in this section is an attempt to gather a list of potentially observable meteor streams. Where no ZHR value is provided it means that visual observation is very difficult, and rates are extremely low. Those streams in the gray bars are either daylight-only streams or those primarily studied only by radio techniques. Observation of weak streams by radio is most effective with radar equipment, but group studies with forward scatter techniques may prove useful.

The radio meteor data presented are the results of forward scatter observations supplied by Andy Smith G7IZU, a UK-based observer. Comparison of activity from month to month should be done with care. Notice at the right hand side of the diagram is a color scale. The calibration of this color scale changes from month to month. The red end of the spectrum denotes peak hourly rates for that month only and will change from month to month. The consequence of this for a month containing major activity such as January is that weaker activity is subdued somewhat. The diagram is known as a colorgram and is a format used by the informal group based around the website http://www.rmob.org. Software is available from that site to convert text-based data to colorgrams, and observers are encouraged to submit their results.

January

Name	Dates	Peak date	Radiant R.A. hour (s: minutes)	Radiant Dec. (degrees)	Speed (km/s)	HR	Population index
α Aurigids	Dec 11 – Jan 21	Jan 1	05:08	+35			
α Quadrantids	Jan 1–5	Jan 4	15:20	+49	41	120	2.1
α Velids	Jan 1–15	Jan 5	08:20	47	35	2	
α Geminids	Dec 28 – Jan 28	Jan 8–9	07:12	+32			
α Crucids	Jan 6–28	Jan 15	12:48	−63	50	3	
Jan Draconids	Jan 10–24	Jan 13–16	16:23	+62			
η Craterids	Jan 11–22	Jan 16–17	11:44	−17			
Jan Bootids	Jan 9–18	Jan 16–18	15:04	+44			
δ Cancrids	Jan 1–31	Jan 17	08:40	+20	28	4	
α Hydrids	Jan 5 – Feb 14	Jan 19	08:52	−11	44	2	
η Carinids	Jan 14–27	Jan 21	10:40	−59		2	
Canids	Jan 13–30	Jan 24–25	07:27	+10			
α Leonids	Jan 13 – Feb 13	Jan 24–31	10:24	+9			
α Carinids	Jan 24 – Feb 9	Jan 30	06:20	−54	25	2	

January is generally a quiet month for meteor activity, with the exception of the Quadrantid meteor stream in the first few days of the month. The peak activity occurs around solar longitude 283.2°, falling in the early hours of January 4. The peak is short and sharp – the time to decline from peak rate to half peak rate is only about 4 h. For this reason, visual observations can be completely clouded out for the whole year for a single observer. Cloud has no effect on radio detection. ZHR reaches a peak of around 120 for visual observers (Fig. 3.6).

Attributing a parent comet to the Quadrantid stream has proved difficult, owing to the relatively rapid evolution of the stream caused by perturbations by Jupiter near the aphelion of the particles' orbit. Over the last 2,000 years the stream orbit has oscillated back and forth and for some time did not intersect with the orbit of Earth. Z. Kanuchova and L. Neslusan of the Slovak Academy of Sciences published

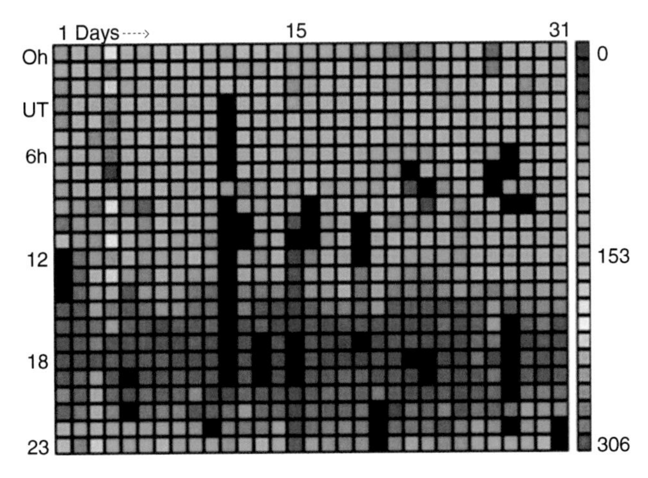

Fig. 3.6. Radio meteor rates for January 2008. Clearly the Quadrantid meteor stream dominates activity in January. *Black squares* indicate times when no data was recorded. (Image courtesy of Andy Smith G7IZU).

a paper in 2007 suggesting the possibility of two objects being parents to the Quadrantids: Comet 96P/Machholz and asteroid 2003 EH1. Uncertainties in the long-term orbital elements and non-gravitational effects meant it was not possible to determine which of the objects is in fact the true parent. Evidence suggested that comet Machholz and asteroid 2003 EH1 may have been a single comet that split, although this is almost impossible to prove.

February

Name	Dates	Peak date	Radiant R.A. (hours: minutes)	Radiant Dec. (degrees)	Speed (km/s)	ZHR	Population index
Capricornids Sagittariids	Jan 13 Feb 28	Feb 1	19:56	−15		15?	
δ Velids	Jan 22 – Feb 21	Feb 05	08:44	−52	35	1	
Aurigids	Jan 31 – Feb 23	Feb 5–10	~04:56	~+42			
α Centaurids	Jan 28 – Feb 21	Feb 07	14:00	−59	56	6	2.0
β Centaurids	Feb 2–25	Feb 8–9	13:52	−58			
o Centaurids	Jan 31 – Feb 19	Feb 11	11:48	−56	51	2	
χ Capricornids	Jan 29 – Feb 28	Feb 13	21:04	+21	27		
Centaurids	Jan 23 – Mar 12	Feb 21	14:00	−41	60	4	
δ Leonids	Feb 15 – Mar 10	Feb 24	11:12	+16	23	2	3.0
σ Leonids	Feb 9 – Mar 13	Feb 25–26	11:17	+14			

There are no major showers in February, but some sources refer to the February to April period as being fireball season. The Capricornids/Sagittariids is a daytime stream only detectable by radio means. Forward scatter detection is challenging, though. There is some recent evidence from the group observations submitted to the International Meteor Organization there is a peak around February 1 at 9:00 UT (Fig. 3.7).

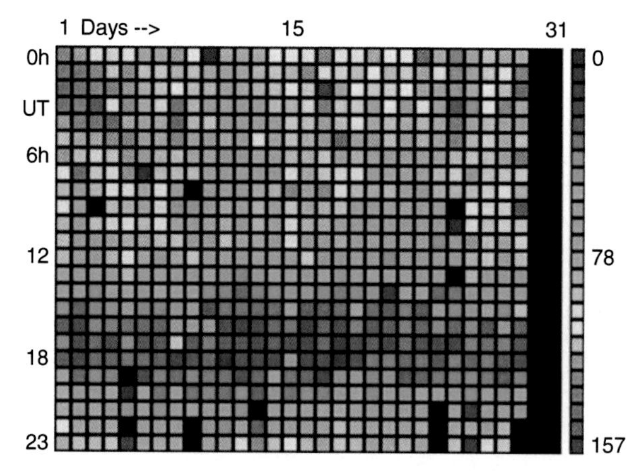

Fig. 3.7. Radio meteor activity for February 2008. Compare this data with the preceding month of January, and it appears there is much more going on in February. This is in fact not true. There is no major activity. The color scale is somewhat stretched; the peak rates this month of 157 per hour are almost half the January peak of 306. (Image courtesy of Andy Smith G7IZU).

The χ Capricornids are also a daytime stream, with evidence of peak activity occurring around 10:00 UT on February 13; once again detection by forward scatter will be challenging, so try it!

There is a question whether the Aurigids are still observable. Historically it was known for bright fireballs, with average magnitude of stream members in the range 3–5. In recent years radiant determination has proved difficult due to low observed rates.

March

Name	Dates	Peak date	Radiant R.A. (hours: minutes)	Radiant Dec. (degrees)	Speed (km/s)	ZHR	Population index
ρ Leonids	Feb 13 – Mar 13	Mar 1–4	10:44	+7			
π Virginids	Feb 13 – Apr 8	Mar 3–9	12:08	+3		2.5	
Leonids/Ursids	Mar 18 – Apr 7	Mar 10–11	11:06	+28			
γ Normids	Feb 25 – Mar 22	Mar 13	16:36	−51	56	8	2.4
March Aquarids	?	Mar 15–18	22:32	−8			
δ Mensids	Mar 14–21	Mar 18–19	03:40	−80		1	
η Virginids	Feb 24 – Mar 27	Mar 18–19	12:20	+3		1	
β Leonids	Feb 14 – Apr 25	Mar 19–21	11:48	+11		3	
θ Virginids	Mar 10 – Apr 21	Mar 20–21	12:56	−2		1	
δ Pavonids	Mar 11 – Apr 16	Mar 30	13:00	−65	31	5	

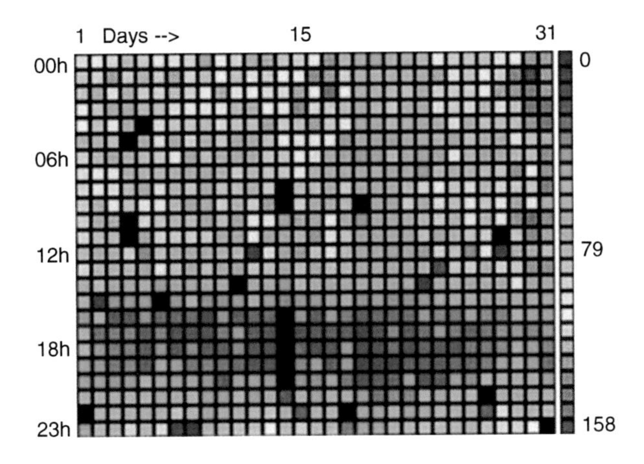

Fig. 3.8. Radio meteor data for March 2008. (Image courtesy of Andy Smith G7IZU).

March is another weak month for meteor observing generally. The Virginid streams are all interrelated and are often referred to as the Virginid complex, which occurs over a 4-month period from February to May. Activity is weak visually, and the extended period of activity makes it difficult to isolate stream activity from sporadic background in forward scatter radio studies (Fig. 3.8).

The March Aquarids is another daylight stream that is little understood and requires more observation.

April

Name	Dates	Peak date	Radiant R.A. (hours: minutes)	Radiant Dec. (degrees)	Speed (km/s)	ZHR	Population index
τ Draconids	Mar 13 – Apr 17	Apr 1–2	19:00	+69			
Librids	Mar 11 – May 5	Apr 17–18	15:28	−16			
δ Pavonids	Mar 21 – Apr 8	Apr 5–6	20:12	−63			
α Virginids	Mar 10 – May 6	Apr 7–18	13:36	−11			
γ Virginids	Apr 5–21	Apr 14–15	12:20	−1			
Apr. Ursids	Mar 18 – May 9	Apr 19–20	09:56	+55			
Apr. Piscids	Apr 8–29	Apr 20–21	00:28	+5			
Lyrids	Apr 15–28	Apr 22	18:04	+34	49	15	2.1
π Puppids	Apr 15–28	Apr 23	07:20	−45	18	Var	2
α Bootids	Apr 14 – May 12	Apr 28	14:32	+19	20	2	
μ Virginids	Apr 1 – May 12	Apr 29	15:08	−07	30	2	

April brings the second major meteor stream of the year, the Lyrids, although peak ZHR only reaches about 15 and constitutes the lower limit of what can be classed major activity (Fig. 3.9).

The Alpha Virginid stream is the strongest branch of this long-duration group of showers, reaching ZHR values of 5–10 in the period April 5–18.

The April Piscid stream is only observable by radio methods, as it is another daytime stream. It was first discovered in a radar survey in 1960 by B.L. Kashcheyev and V.N. Lebedinets in the USSR. It was later observed in another three radar surveys, all yielding inconsistent hourly rates. This leads to the question of whether it undergoes periodic variation in activity.

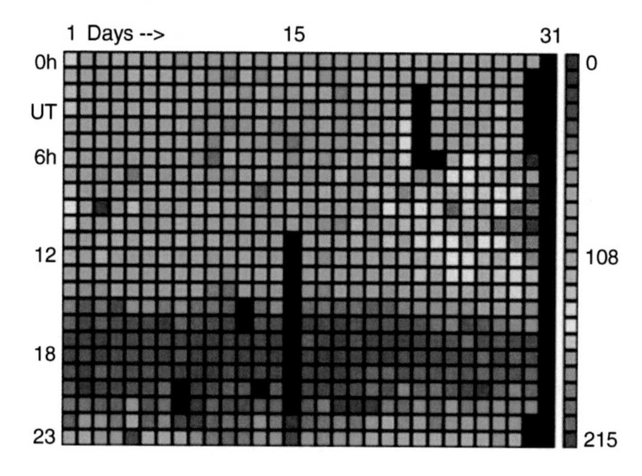

Fig. 3.9. Radio meteor data for April 2008. Meteor activity is low in early April, but significantly picks up towards the end of the month and into June. (Image courtesy of Andy Smith G7IZU).

May

Name	Dates	Peak date	Radiant R.A. (hours: minutes)	Radiant Dec. (degrees)	Speed (km/s)	ZHR	Population index
ω Capricornids	Apr 19 – May 15	May 02	21:00	−22	50	2	
η Aquariids	Apr 19 – May 28	May 06	22:32	−01	66	60	2.4
May Librids	May 1–9	May 6–7	15:32	−18		3	
η Lyrids	May 3–12	May 09	19:08	+44	44	3	3.0
ε Arietids	Apr 25 – May 27	May 9–10	02:56	+21			
May Piscids	May 4–27	May 12–13	00:52	+22			
α Scorpiids	May 1–31	May 16	16:12	−21	35	5	
β Corona Austrinids	Apr 23 – May 30	May 16	18:56	−40	45	3	
May Arietids	May 4 – Jun 6	May 16–17	02:28	+18			
o Cetids	May 7 – Jun 9	May 14–25	01:52	−3			
Southern May Ophiuchids	Apr 21 – Jun 4	May 13–18	16:48	−23			
ε Aquilids	May 4–27	May 17–18	16:24	+13			
Northern May Ophiuchids	Apr 8 – Jun 16	May 18–19	16:52	−15		2	

The η Aquarids produce major activity peaking around May 6, although this is a challenging meteor stream for visual observers. The radiant does not rise until about an hour before dawn for mid-latitudes, and the situation is even worse for northern or southern latitudes. The bulk of the activity occurs after dawn. This clearly shows as a hot spot in the May colorgram for forward scatter radio observers (Fig. 3.10).

May is a great month for daytime meteor activity. There are four daytime only streams: the ε Arietids, May Arietids, o Cetids, and May Piscids. The o Cetids can produce as many as 18 radio echoes per hour near peak time.

The ε Aquilids were discovered by radar surveys in the 1960s. It is not known to have been observed visually.

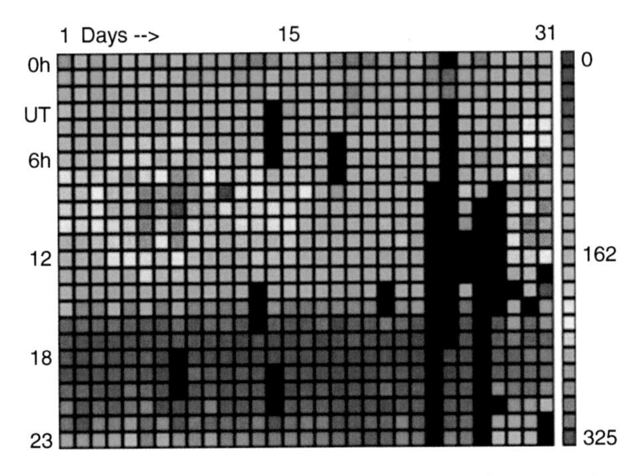

Fig. 3.10. Radio meteor data for May 2008. May sees the peak of one of Halley's Comet derived meteor streams the η Aquarids. There is also lots of daytime activity. (Image courtesy of Andy Smith G7IZU).

June

Name	Dates	Peak date	Radiant R.A. (hours: minutes)	Radiant Dec. (degrees)	Speed (km/s)	ZHR	Population index
χ Scorpids	May 6 – Jul 2	May 28 – Jun 5	16:20	−12			
ω Scorpiids	May 23 – Jun 15	Jun 2	15:56	−20	21	5	
Arietids	May 22 – Jul 2	Jun 7	02:56	+24	38	60	
τ Herculids	May 19 – Jun 19	Jun 9	15:44	+41			
ζ Perseids	May 20 – Jul 5	Jun 13	04:12	+26		40	
June Lyrids	Jun 10–21	Jun 15	18:20	+35		Var 1.3–3.5	
June Aquilids	Jun 2 – Jul 2	Jun 16–17	19:48	−7			
Scorpids Sagittarids	Jun 1 – Jul 15	Jun 19	18:16	−23	30	5	
Ophiuchids	May 19 – Jul 2	Jun 20	17:32	−20		6	
June Scutids	Jun 2 – Jul 29	Jun 27	18:32	−4		2	
τ Cetids	Jun 18 – Jul 4	Jun 27	01:36	−12	66	4	
June Bootids	Jun 28 – Jul 5	Jun 28	14:36	+49	14	Var	2.2
τ Aquariids	Jun 19 – Jul 5	Jun 28	22:48	−12	63	7	
θ Ophiuchids	Jun 4 – Jul 15	Jun 29	16:36	−15	29	2	
β Taurids	Jun 5 – Jul 18	Jun 29	05:17	+21		25	

The Arietids is the best known and the strongest of the daytime only meteor streams, rivaling the η Aquarids of May, which is virtually a daytime stream. This is followed up by the ζ Perseids, giving a significant daytime morning response on forward scatter radio results in the first half of the month. The latter half of the month is dominated by the daytime activity of the β Taurids (Fig. 3.11).

June is a difficult month for many visual observers, especially in high northern latitudes, due to the short nights that never reach full darkness. This is compensated for by the richness of the daytime activity.

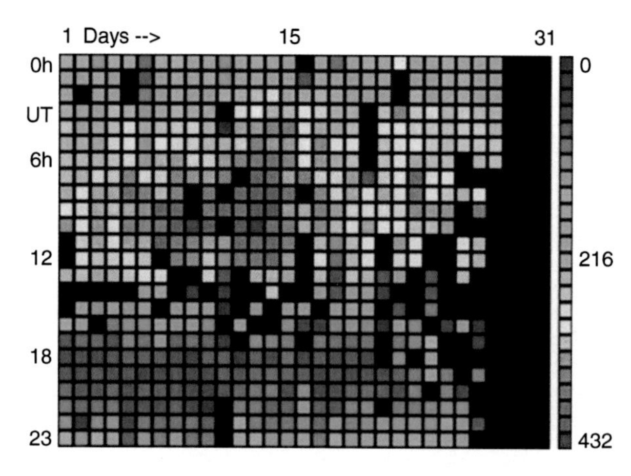

Fig. 3.11. Radio meteor data for June 2008 The strongest daytime activity occurs in June, with three very active daytime showers. It is also a difficult month for forward scatter when Sporadic E interference is common. (Image courtesy Andy Smith G7IZU).

Meteors and Meteor Streams

July

Name	Dates	Peak date	Radiant R.A. (hours: minutes)	Radiant Dec. (degrees)	Speed (km/s)	ZHR	Population index
July Pegasids	Jul 7–13	Jul 10	22:40	+15	70	3	
July Phoenicids	Jul 10–16	Jul 13	02:08	–48	47	Var	
τ Capricornids	Jun 27 – Jul 29	Jul 12–13	20:42	–15			
α Lyrids	Jul 9–20	Jul 14–15	18:40	+38		1	
o Draconids	Jul 6–28	Jul 17–18	18:58	+61			
α Cygnids	Jul 11–30	Jul 18	20:20	+47	37	2	
σ Capricornids	Jul 15 – Aug 11	Jul 20	20:28	–15	30	5	
Piscis Austrinids	Jul 15 – Aug 10	Jul 28	22:44	–30	35	5	3.2
Southern δ Aquariids	Jul 12 – Aug 19	Jul 28	22:36	–16	41	10	3.2
α Capricornids	Jul 3 – Aug 15	Jul 30	20:28	–10	23	4	2.5

The main shower of July is the Southern δ Aquarids, peaking towards the end of the month but not quite a major stream of the year. Once again Sporadic E interference in July is common, making forward scatter work impossible (Fig. 3.12).

The α Lyrid stream is an interesting one; the ZHR barely reaches 2 for naked-eye work, but it is known as a strongly active telescopic shower, where rates of 18 per hour can be seen with binoculars. This should prove a good target for forward scatter study and should yield measurable underdense activity.

Visual records of the o Draconids are weak, and although Sekanina recorded it in his 1968–1969 session of the Radio Meteor Project, it seems that little observation of it has occurred since then.

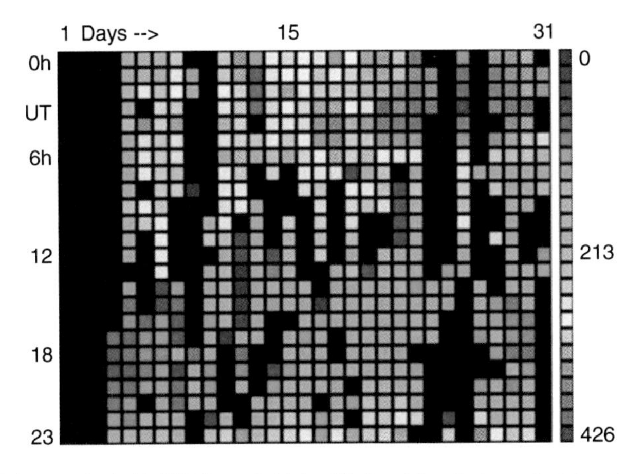

Fig. 3.12. Radio meteor data for July 2008. Once again July is a difficult month for radio observers due to frequent periods of Sporadic E interference. (Image courtesy Andy Smith G7IZU).

August

Name	Dates	Peak date	Radiant R.A. (hours: minutes)	Radiant Dec. (degrees)	Speed (km/s)	ZHR	Population index
α Capricornids	Jul 15 – Sep 11	Aug 1–2	20:26	−8		8	
Southern ι Aquariids	Jul 25 – Aug 15	Aug 04	22:16	−15	34	2	
Northern δ Aquariids	Jul 15 – Aug 25	Aug 08	22:20	−05	42	4	
Perseids	Jul 17 – Aug 24	Aug 12	03:04	+58	59	90	2.6
α Ursa Majorids	Aug 9–30	Aug 13–14	11:00	+63		4	
κ Cygnids	Aug 3–25	Aug 17	19:04	+59	25	3	3.0
Northern ι Aquariids	Aug 11–31	Aug 20	21:48	−06	31	3	
π Eridanids	Aug 20 – Sep 15	Aug 25	03:28	−15	59	4	
γ Leonids	Aug 14 – Sep 12	Aug 25–26					
γ Doradids	Aug 19 – Sep 6	Aug 28	04:36	−50	41	5	

The Perseids dominate the month of August on sheer hourly rates, although there is a lot of other interesting activity going on at the same time. Perseids always give a good show to both visual and radio observers (Fig. 3.13).

This is a great time of year for meteor observing. The α Capricorndid stream produces some spectacular bright and very slow meteors. Although it is always difficult to attribute forward scatter radio echoes to a particular stream, the α Capricorndids should generate some good overdense radio echoes.

Similarly, the κ Cygnids produce a low rate of relatively bright meteors and has always been of interest to radio amateur astronomers. It is all too easy to ignore it, due to the observers' preoccupation with the Perseids. Take time out to identify these weaker August streams visually, and they will reward you.

The α Ursa Majorids is not well studied; although it is not a significant shower to the naked eye there is some evidence it is much more active telescopically and to radar instruments.

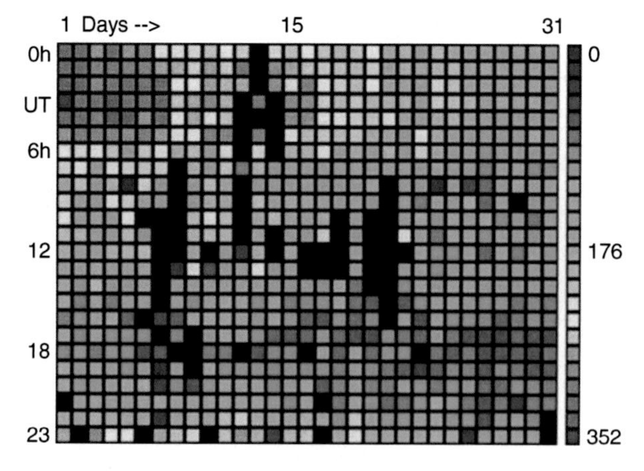

Fig. 3.13. Radio meteor data for August 2008. One of the best known showers the Perseids, peak around August 13. It always puts on a good show for visual and radio observers. (Image courtesy of Andy Smith G7IZU).

The γ Leonid stream is a daylight shower. It was seen for the first time in the 1960s in two radar surveys, and its very weak activity even in radar studies has meant no accurate rate determinations have been made. It is unlikely that forward scatter results will be able to detect it from the background sporadic activity.

September

Name	Dates	Peak date	Radiant R.A. (hours: minutes)	Radiant Dec. (degrees)	Speed (km/s)	ZHR	Population index
α Aurigids	Aug 25 – Sep 8	Sep 01	05:36	+42	66	7	2.6
γ Aquarids	Sep 1–14	Sep 7				2	
September Perseids	Sep 5 – Oct 10	Sep 08	04:00	+47	64	6	2.9
Aries-Triangulids	Sep 9–16	Sep 12	02:00	+29	35	3	
Piscids	Sep 1–30	Sep 20	00:32	+00	26	3	

The α Aurigids, although fairly weak in activity, is the strongest meteor stream of the month, certainly for visual work (Fig. 3.14).

The September Perseids appear as a fairly new stream on the annual calendar, but in fact activity now attributed to this radiant was formerly classed as part of the δ Aurigid stream.

The Aries Triangulids is poorly understood and was first observed by two experienced observers in the 1980s, Gary Kronk and George Gliba.

The Piscid stream is very diffuse, with an ill defined peak of activity in the range September 11 to around September 20. There are suggestions of more than one radiant from this stream, but there are too few observations to be sure.

September is one of those odd months. It suffers from too few active observers, perhaps because at that time people realize the summer is over and winter is coming. The children are back at school, and the weather is getting colder, especially at night. The moral of this story is – get out and observe!

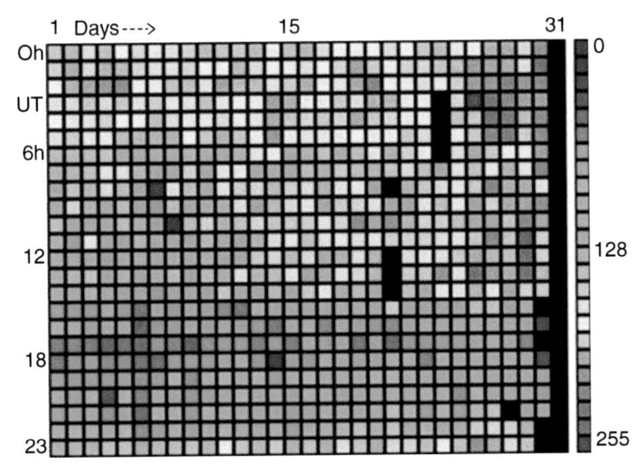

Fig. 3.14. Radio meteor data for September 2008. September is traditionally a poorly observed month by visual workers. Visual streams are weak and complex. (Image courtesy of Andy Smith G7IZU).

October

Name	Dates	Peak date	Radiant R.A. (hours: minutes)	Radiant Dec. (degrees)	Speed (km/s)	ZHR	Population index
Sextantids	Sep 24 – Oct 9	Oct 1–4	10:12	−2			
October Arietids	Oct 1–31	Oct 08	02:08	+08	28	5	
Giacobinids/Draconids	Oct 6–10	Oct 08	17:28	+54	20	Var	2.6
δ Aurigids	Sep 22 – Oct 23	Oct 10	05:40	+52	64	6	
ε Geminids	Oct 14–27	Oct 18	06:56	+27	71	2	3.0
Orionids	Oct 2 – Nov 7	Oct 21	06:20	+16	66	20	2.5
Leo Minorids	Oct 21–23	Oct 22	10:48	+37	62	2	3.0

The month opens with a weak diffuse daylight meteor stream, the Sextantids. However its discovery stems from a radio survey by A.A. Weiss in 1957, when a maximum rate of 30 per hour were recorded. The stream may be periodic in nature, undergoing regular outbursts as Earth encounters the stream every 4–5 years or so (Fig. 3.15).

Another periodic stream is the Giacobinids, also known as the Draconids. Most of the year's activity is low, but several times over the last century outbursts of hundreds or even thousands per hour have been seen. The outburst years were 1933, 1946, 1952, 1985, 1998, and 2005. Rates become high when the parent comet Giacobini-Zinner returns. However, not all comet return have yielded high rates; for example, 1972 was a big disappointment. In more recent times, activity rates in 1998 and 2005 reached as high as 150 per hour.

The only regular major activity in October is the Orionid stream. This is the second of the Halley's Comet dust streams, the first being the η Aquarids. The reason for this dual encounter is that the orbit of the comet is lying almost in the same plane as that of Earth, so we encounter it twice, 6 months apart.

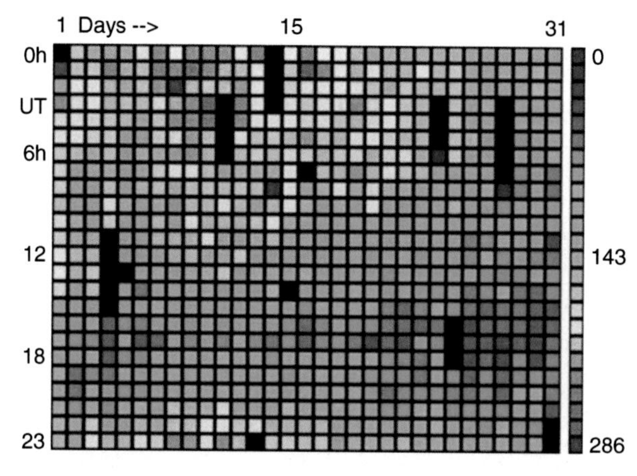

Fig. 3.15. Radio meteor data for October 2008. October sees the return of the Halley's Comet dust stream in the form of the Orionid shower. (Image courtesy of Andy Smith G7IZU).

November

Name	Dates	Peak date	Radiant R.A. (hours: minutes)	Radiant Dec. (degrees)	Speed (km/s)	ZHR	Population index
Southern Taurids	Nov 1–25	Nov 05	03:28	+13	27	5	2.3
δ Eridanids	Nov 6–29	Nov 10	03:52	−09	31	2	
Northern Taurids	Nov 1–25	Nov 12	03:52	+22	29	5	2.3
ζ Puppids	Nov 2 – Dec 20	Nov 13	07:48	−42	41	3	
Leonids	Nov 14–21	Nov 17	10:12	+22	71	Var	2.5
α Monocerotids	Nov 15–25	Nov 21	07:20	+03	60	Var	2.4

The Leonid stream dominates the thoughts of any meteor observers in November, even though most years rates can barely reach the 15 per hour to warrant its major status. Roughly every 33 years shortly after the passage of the parent comet Tempel-Tuttle the peak observable rates increase considerably. It is always difficult to predict which returns may produce storm levels of activity, though. Many people remember the 1999/2000 return of the comet as a period. Predictions of Earth's passage through a filament stream in 1999 had proven remarkably accurate. Video data agreed within 6 min the accepted peak time of the stream for that year from collected group observations (Fig. 3.16).

The α Monocerotids is another of the variable rate streams that undergoes regular outbursts of up to 100 per hour. The details of this stream, however, are still uncertain, due to the normally low rates of faint meteors.

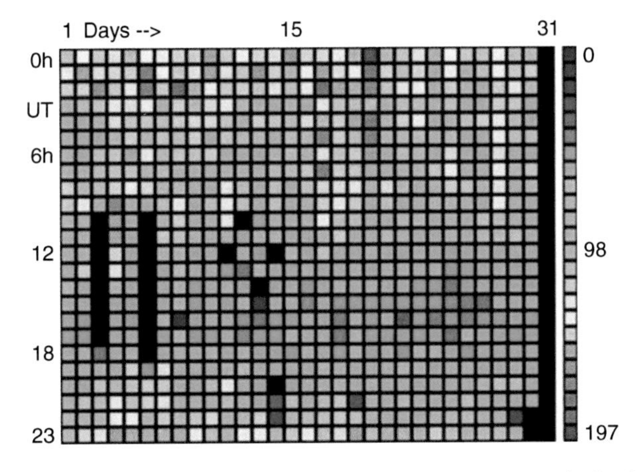

Fig. 3.16. Radio meteor data for November 2008. The only classical major stream of the month is the Leonids, although in most years it fails to deliver major hourly rates. (Image courtesy of Andy Smith G7IZU).

Meteors and Meteor Streams

December

Name	Dates	Peak date	Radiant R.A. (hours: minutes)	Radiant Dec. (degrees)	Speed (km/s)	ZHR	Population index
χ Orionids	Nov 25 – Dec 31	Dec 02	05:28	+23	28	3	
α Pupids	Nov 17 – Dec 9	Dec 2–5	08:32	−45			
σ Hydrids	Dec 4–15	Dec 11–12	08:28	+2		3	
Phoenicids	Nov 28 – Dec 9	Dec 06	01:12	−53	18	Var	2.8
Monocerotids	Nov 27 – Dec 17	Dec 09	06:48	+08	43	3	3.0
11 Canis Minorids	Dec 4–15	Dec 10–11	07:48	+13			
Northern χ Orionids	Nov 16 – Dec 16	Dec 10–11	05:28	+23		2	
Southern χ Orionids	Dec 2–18	Dec 10–11	05:52	+20			
Dec. Monocerotids	Nov 9 – Dec 18	Dec 11–12	06:44	+10		1	
σ Hydrids	Dec 3–15	Dec 12	08:28	+02	58	2	3.0
Puppid-velids	Dec 2–16	Dec 12	09:00	−46	40	4	2.9
Geminids	Dec 7–17	Dec 14	07:28	+33	35	120	2.6
Ursids	Dec 17–26	Dec 22	14:28	+76	33	10	3.0
Coma Berenicids	Dec 12 – Jan 23	Dec 30	11:40	+25	65	5	3.0

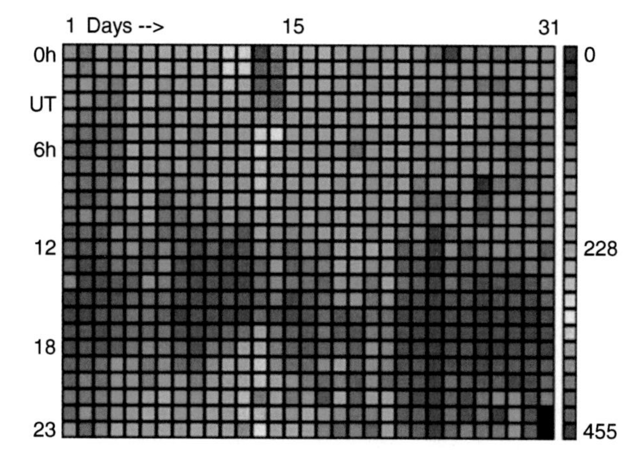

Fig. 3.17. Radio meteor data for December 2008. The Geminid stream dominates the December meteor calendar and is certainly one of the great annual showers to observe. (Image courtesy of Andy Smith G7IZU).

The December meteor activity is dominated by the Geminids, which is probably the best of the annual showers. Activity rates are high shortly after darkness, and the stream puts on a good show from early evening to the early hours of the next day on the peak evening (around December 14). All of the other well known high rate showers are best observed after midnight. Although there are quite a number of minor showers active this month, you can see from the colorgram the Geminids dominate the radio results, too. Similar to January the colorgram technique of displaying meteor radio data does suppress the impact of minor activity for the month (Fig. 3.17).

The Christmas festivities, poor weather, and the cold temperatures of December do tend to reduce the amount of data gathered by visual observers of the late December streams, particularly the Coma Berenicids and the Ursids. Of course, the great advantage of radio observation is that it's completely automated.

The 11 Canis Minorids were first seen in 1964 by Keith B. Hindley, a very active meteor observer of the British Astronomical Association. He was using a short-focus rich-field 5-in. telescope to observe the Geminids telescopically, in order to better determine the Geminid radiant. However, he observed five meteors of magnitude between +6 and +12.

This stream may not be detectable by naked-eye observation. This, of course, lends itself to radio studies, which should produce underdense activity. Once again, though, forward scatter observations make it difficult to attribute echo events to particular streams, and the build up to the Geminids will likely hide its activity in the general background.

The December Monocerotids are extremely weak as a visual meteor stream; however, radar studies show the stream makes a regular annual return, so again a well set up sensitive forward scatter system should detect them, although again the buildup of Geminid activity will likely mask out the Monocerotids.

Chapter 4

Beyond the Solar System

Before we discuss specific details about radio objects outside our Solar System it is appropriate to introduce some fundamental concepts of radio astronomy and the basic scientific measurements gathered by radio telescopes. In this section we will introduce the units of measurement used to compare the relative strengths of radio objects in the sky.

Brightness and Flux Density

It is very important to understand the concepts of brightness and flux density. The ideas are simple enough but easily confused with each other. These concepts are not unique to the radio spectrum, of course. They apply for all electromagnetic radiation (Fig. 4.1).

Let's consider the simplest case where radiation travels through empty space. It is not absorbed or scattered along the way, and there are no extra emission sources on route.

The ray-optics approximation considers the energy to be flowing in straight lines. This approximation is valid only if the source is physically much larger than the wavelength of the radiation. Clearly this is true for astronomical sources such as planets, stars, and nebulae.

Now consider the Sun, which appears to have a nearly uniform "brightness" distribution across its face. If a camera is used to take its picture, the exposure used would not vary whether it was taken from Venus, Earth, or Mars. Only the angular diameter of the Sun would vary. The Venus photograph would not be overexposed, and the Mars photograph would not be underexposed. The number of photons falling on the film per unit area per unit time per unit solid angle remains constant, regardless of distance. In other words the intrinsic brightness of the Sun does not change with distance.

The number of photons falling on the film per unit area per unit time does decrease with distance. Therefore brightness or specific intensity is a measure of the energy received per unit area per unit time per unit solid angle. While the flux is the energy received per unit area per unit time, specific intensity is constant. Flux reduces

J. Lashley, *The Radio Sky and How to Observe It*, Astronomers' Observing Guides,
DOI 10.1007/978-1-4419-0883-4_4, © Springer Science+Business Media, LLC 2010

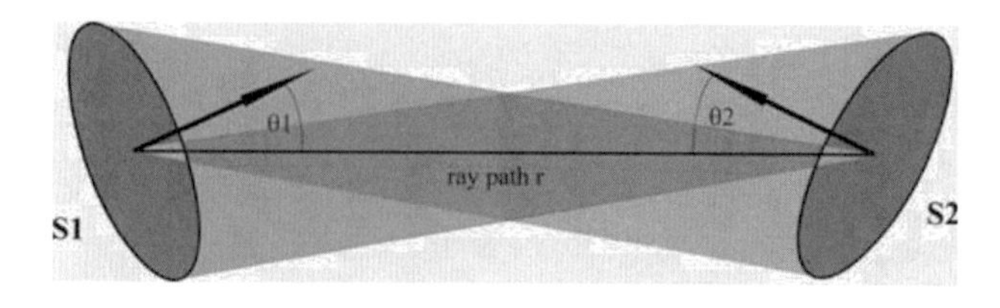

Fig. 4.1. Specific intensity. Our normal thoughts of brightness relate to the intensity of a source per unit area. This clearly decreases with distance from the source. However, specific intensity is the power per unit area per unit solid angle Θ per hertz. This is constant at any distance from the source.

in inverse proportion to the square of the distance from the source. Other names for "brightness" are spectral intensity, spectral brightness, and spectral radiance.

Specific intensity is therefore conserved (constant) along any ray in empty space (no absorption or scattering) in which case the brightness can be considered as energy flowing out of the source, or as energy flowing into our detector.

For discrete sources (those that subtend a well-defined solid angle) the spectral power received by the antenna of unit area is called the flux density. If a source is unresolved by the telescope, then its flux density can be measured, but its specific intensity cannot. If a source is much larger than the point source response and is resolved, the spectral intensity can be measured directly, but the flux density must be calculated by integrating the spectral intensity over the source's solid angle.

The unit of flux density is $Wm^{-2}\ Hz^{-1}$. A value of 1 $Wm^{-2}\ Hz^{-1}$ is a large figure in radio astronomy, so the Jansky was introduced, where 1 Jansky = $10^{-26}\ Wm^{-2}\ Hz^{-1}$.

For most objects observed with small radio telescopes, including the Sun, we can only determine the flux density.

Often in radio astronomy the brightness, or specific intensity, of an object is given as a temperature. Quantum mechanics provides a relationship between received power and wavelength that is accurate over the whole of the electromagnetic spectrum in the Planck equation. However it is cumbersome to work with, whereas for the restricted bandwidths of the radio spectrum the Rayleigh–Jeans law is a good approximation. The Rayleigh–Jeans formula relates power to temperature and wavelength:

$$P = \frac{2\pi kT}{\lambda^2} Wm^{-2} Hz^{-1}$$

where P is the power emitted by 1 m^2 of a blackbody surface whose temperature is T kelvin, over a bandwidth of 1 Hz at the wavelength of λ. K is the Boltzmann constant

We know, however, that the flux density is attenuated by the inverse square law as the radiation travels outward from the source. The amount of attenuation is:

$$attenuation\ due\ to\ distance = \left(\frac{R}{d}\right)^2$$

where R is the radius of the source and d its distance from us.

We can then modify the Rayleigh–Jeans equation to relate our received flux density to the temperature and wavelength like this:

$$S = \frac{2\pi kT}{\lambda^2}\left(\frac{R}{d}\right)^2 Wm^{-2} Hz^{-1}$$

where S is our measured flux density.

Continuum Emission

Fundamentally, radio emission falls into two broad classes, thermal and non-thermal. Thermal emission occurs from a system whose population state is associated with the Maxwell–Boltzmann velocity distribution, which is dependent on the kinetic temperature T. The intensity of the emission may depend on various parameters, but it will certainly be a function of the temperature. In a heated gas or plasma the motion of the particles is random, and emitted radiation therefore exhibits no specific polarization. In practice temperatures greater than 10^8 K are rarely encountered, so if a derived temperature is higher than 10^8 K the chances are it is of non-thermal origin. Non-thermal radiation is everything else and is caused by charged particles being accelerated (or decelerated) in the presence of an electric or magnetic fields, or when particles collide with other particles whose velocity distribution does not match the Maxwellian profile.

Thermal Bremsstrahlung Spectrum

Bremsstrahlung is a word derived from the German language for "braking radiation." It occurs when a free electron travels close to but does not combine with a positively charged atomic nucleus or ion. Another name given to Bremsstrahlung radiation is free-free emission, referring to the fact that the electron is free before and after the emission of a photon (Fig. 4.2).

The emission occurs when the electron encounters an electrostatic force, deflecting it in its in path and therefore emitting a radio photon. Because energy must be conserved the result is to slow down the electron, hence the name "braking radiation." In nature the deflection angle is usually small, and the associated change in velocity negligible, resulting in very low energy photon emission in the radio spectrum.

The spectrum of Bremsstrahlung radiation will have a sharp cutoff at upper and lower ends. At the low frequency end the emission is from electrons that approach the nuclei closely, and the cutoff occurs because closer encounters involve election capture. The high-frequency emission is limited by the temperature of the gas, so if this cutoff can be observed it is all the information required to provide a measure of the gas temperature. Note here the upper cutoff may be outside of the radio spectral window depending on cloud temperature.

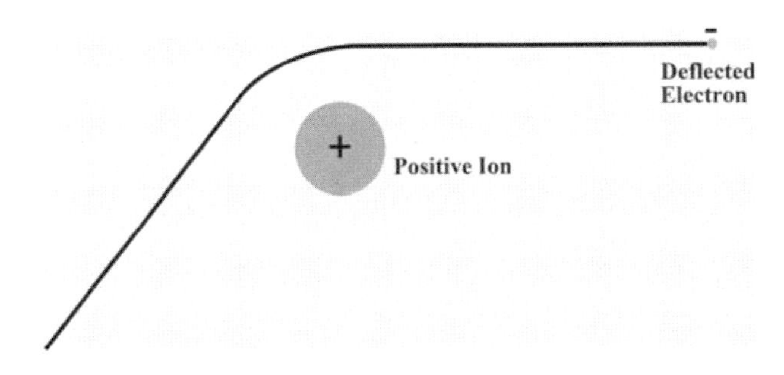

Fig. 4.2. A fast-moving electron passing close to a slow-moving heavy ion is deflected and loses a small amount of energy in the form of a radio photon.

Bremsstrahlung is often the dominant process for radio emission from ionized HII clouds and planetary nebulae. Study of the Bremsstrahlung spectrum can yield information on the density of the cloud. Radio studies are by far the best way to obtain this information. Radio waves are not scattered or attenuated by the interstellar medium (ISM) because the wavelength is much greater than the average particle size of the ISM. Scattering or attenuation of light often requires uncertain corrections to be applied to observations. Once a measure of the density is known the mass of the cloud can be determined from its volume. The assumption for cloud volume is, it is about as deep as it is wide.

Cyclotron and Synchrotron Radiation

In a similar way to Bremsstrahlung, emission of energy occurs from electrons deflected in a magnetic field. In the synchrotron case, the electrons have relativistic velocities (close to the speed of light), and for cyclotron emission the electron velocity is much less than that of light. Occasionally this process is referred to as magneto-Bremsstrahlung.

The electron traveling in a magnetic field experiences a Lorentz force, which is perpendicular to both the velocity and magnetic field vectors. The velocity vector can be at any angle to the magnetic field and is known as the pitch angle (Fig. 4.3).

The result of an arbitrary pitch angle is that the electron will move parallel to the magnetic field in a spiral path. If the field line curves, then the spiral path curves with it. The electron is trapped in the field. In the special case of a zero pitch angle the electron will move in a circular path around the magnetic field. The other special case of a 90° pitch angle means the electron is unaffected by the field.

Consider first the cyclotron case. Cyclotron radiation is emitted at the gyrofrequency. When viewed from a direction along the field line, the electric vector of the emitted radiation will be seen to rotate, therefore giving rise to circularly polarized emission. When viewed from the side, however, the motion of the electron appears to oscillate from side to side, and therefore the polarization will be linear. For all viewing angles in between the emission will appear to have elliptical polarization. For this reason polarized radio emission is a definite indication of a magnetic field within the source.

Note that the gyrofrequency is independent of the electron velocity. The radius of the circular or spiral motion (the gyroradius) is defined only by the strength of the magnetic field, so the gyrofrequency is also only dependent on the magnetic field strength. The gyrofrequency has a very simple relation to the field strength given by the following formula:

$$f = 2.8B$$

where f is the gyrofrequency in hertz and B is the magnetic field strength in gauss.

Clearly, if the cyclotron frequency can be measured, the strength of the field can be immediately determined. However, plasmas exhibit a critical frequency known

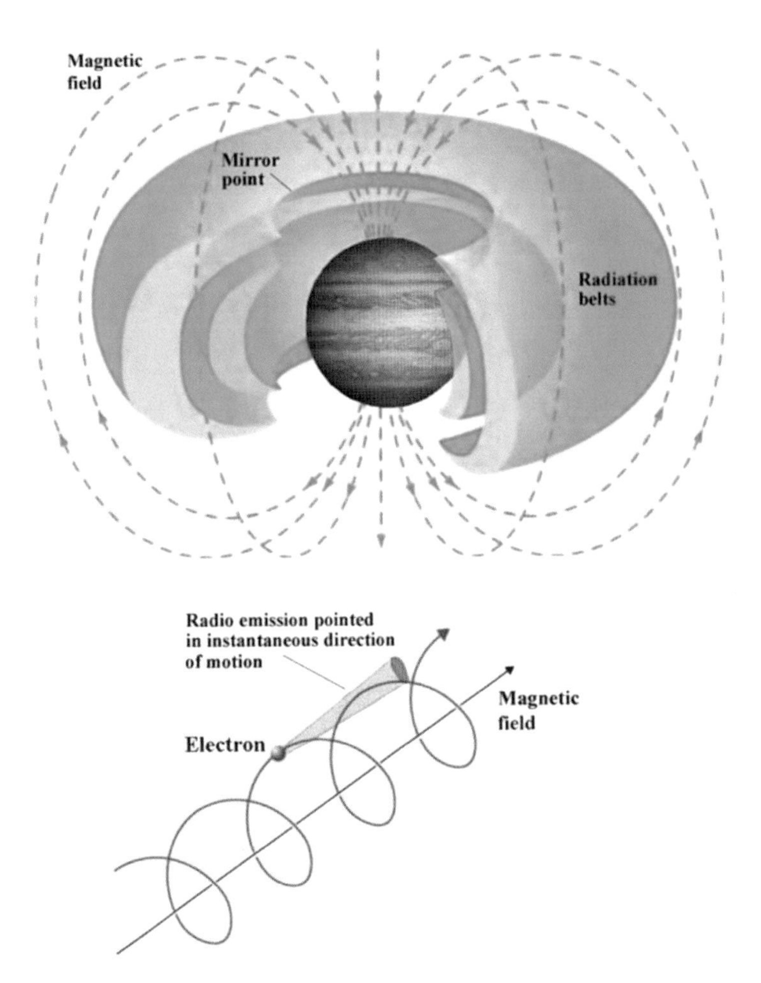

Fig. 4.3. The cyclotron/synchrotron emission process from trapped electrons in a magnetic field.

as the plasma frequency, such that a wave emitted within it must have a frequency higher than the plasma frequency in order to escape, or else it will be reflected. On top of this remember that Earth's ionosphere is also a plasma, and therefore the cyclotron emission has to have a frequency higher than the ionospheric plasma frequency to be observable by ground-based stations.

As an example the warm ionized medium (WIM) of the interstellar regions of our Milky Way Galaxy will emit cyclotron radiation. If the magnetic field strength here is 3 μG and the temperature is 10^4 K the frequency of emission will be 8.4 Hz. Not only will this not penetrate our ionosphere, it will not even escape the WIM because its plasma frequency will be 890 Hz. As we saw in the chapter on Jupiter, though, the strong field of the Jovian environment does allow us to study a good proportion of the Jovian cyclotron emission. The field strength of any object must be greater than 3.5 G or higher in order for cyclotron emission to penetrate our ionosphere, assuming a cutoff of 10 MHz.

Where the magnetic field of the ionized region varies smoothly (such is the case for Jupiter) the cyclotron emission will give a continuum spectrum with a sharp cutoff occurring at the point of maximum field strength.

The Synchrotron Spectrum

Synchrotron emission occurs from electrons traveling at relativistic speeds within a magnetic field. This is a common process occurring throughout the universe. Now, the mass of the electron is increased by a factor γ known as the Lorentz factor. This increases the gyroradius and decreases the gyrofrequency compared to the cyclotron case.

The transformations required between the rest frame of the electron (the frame of reference moving with the electron) and the rest frame of the observer result in the total power being boosted by a factor γ^2. The power as seen by the observer is distributed over a conical pattern whose angle is $1/\gamma^2$. This cone is directed in the path of the instantaneous velocity vector, and only if the cone passes through the observer's line of sight will it be detectable. For high-energy particles the cone of emission is small compared with the gyration frequency, so the energy is pulsed as the cone briefly sweeps past the observer's line of sight. The duration of the pulse defines the highest observable frequency. However, most emissions occur at lower frequencies, which are harmonics of the fundamental gyrofrequency; the result of the significant reduction of the gyrofrequency is that these harmonics are very close together. This means synchrotron emission is essentially continuous and spread over a broad spectral range. Due to the emission occurring significantly above the fundamental gyrofrequency even small magnetic fields produce observable emission, making synchrotron radiation more commonly observed than cyclotron.

Synchrotron radiation is unrelated to temperature. The electron velocities do not follow a Maxwellian distribution. Hence the observed "temperature" can appear excessive, and it is therefore classed as a non-thermal emission.

When analyzing the spectrum of synchrotron sources the results can be a jumble of other emission such as thermal Bremsstrahlung, as would be the case for a supernova remnant. Also in objects such as active galactic nuclei optically thick emission is common, but the spectral detail is complex due to the emission from heavier particles being superimposed on the electron synchrotron spectrum. However if the synchrotron component can be clearly observed, it can be used to measure the magnetic field strength. The rise of the spectrum to the peak value comes from optically thick emission. Optically thick refers to the probability of particle collision resulting in recombination. The depth of the cloud is much greater than the mean free path of the particles, and we cannot see all the way through the cloud. The brightness of the source is therefore independent of the density. From a single measurement of the optically thick spectrum it is possible to determine the strength of the field perpendicular to the observer's line of sight. From a statistical combination of observations an average of the readings will give an approximate value of the overall field strength.

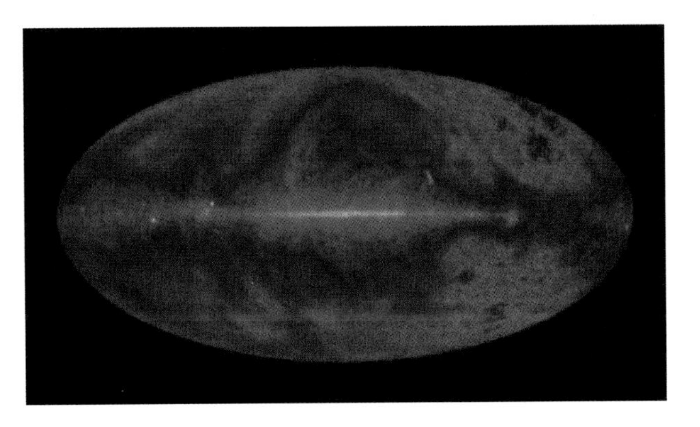

Fig. 4.4. The radio sky at 408 MHz. (C. Haslam et al. Max Planck Institute for Radio Astronomy.) The emission is largely due to synchrotron radiation for our galaxy.

The portion of the synchrotron spectrum beyond the peak, as it falls to minimum, is generated is a power law slope and is characteristic of optically thin emission (where the mean free path is greater than the depth of the cloud, the cloud is transparent). The power law slope is a consequence of the electron energy distribution and means the spectrum is a function of both the particle density and the magnetic field. Unfortunately it is not possible to determine the particle density alone without first knowing the magnetic field strength B. As indicated previously, it is not always possible to isolate the optically thick spectral component to determine B. In such cases, by making some estimates and assumptions it is still possible to extract information on the magnetic field strength and particle energies from observations of the optically thin spectrum, which unlike the optically thick is readily observable.

For the same reason as cyclotron, synchrotron emission is intrinsically polarized. Thermal Bremsstrahlung is never polarized, so the presence of polarization is a good indication that the synchrotron mechanism is the cause (cyclotron emission is rarer and restricted to long wavelengths). However, as always, the reality is not always so clear; in sources of large extent such as the interstellar medium of our Milky Way, the magnetic field can be twisted and more randomly distributed. The consequence is that the combined response across the source region averages out and polarization is mixed. For cases where polarization clearly exists, then synchrotron emission is certain, but the absence of polarization does not rule it out.

The image presented in Fig. 4.4 shows a map of the sky at 408 MHz, consisting of mostly synchrotron emission.

The various sources of radiation in the image include supernovae remnants, shock waves in the interstellar medium, stellar jets, some binary stars, and pulsars; even cosmic rays passing through the interstellar medium can produce interactions generating radiation. The magnetic field pervading the Milky Way may amount to only a few micro gauss, but this is sufficient to produce observable radio radiation.

Extragalactic sources of synchrotron energy include active galactic nuclei powered by super massive black holes, which generate enormous bipolar jets of particles. Two of the brightest examples are Centaurus A and Cygnus A.

Inverse Compton Scattering

Inverse Compton scattering does not tend to produce output in the radio spectrum, but it does potentially reduce the amount of radio photons emitted.

Compton scattering is where a high-energy photon in the X-ray to gamma ray spectrum collides with an electron imparting energy to the electron, and the result increases the wavelength of the initial photon. The inverse process also occurs, where a high-energy electron impacts a low-energy photon, thus decreasing the photon wavelength and pushing it into the visible or higher spectral range. The sources producing Inverse Compton (IC) radiation also produce synchrotron emission, but not all synchrotron sources produce IC.

For sources of a given number of relativistic electrons, if the radiation field is weak and the magnetic field is strong, synchrotron output will dominate. For the inverse case of a weak magnetic field and a strong radiation field, IC radiation will be the dominant process.

Emission Line Radiation

Emission lines exist throughout the electromagnetic spectrum and occur when bound electrons change their quantum energy state from a high level to a lower level, emitting a photon in the process. Because in general the electrons can only hold discrete energy states, then where a large population undergo the same transitions continuously an emission line can be observed. In this section only the radio frequency emission lines will be considered.

Hydrogen is the most common element in the universe, so the significance of rare transitions of electron quantum state can be measured in practice. The highest quantum states for the electron in a hydrogen atom, where $n > 40$, have energy levels that are very close together. Transitions at these high-energy levels will emit radiation in the radio spectrum. In comparison to the emission at lower levels of the atom, which give optical output, these high transitions are rare. It is believed possible that quantum numbers up to 1,600 are possible, where the electron would be placed 0.3 mm from the nucleus!

Line emissions from hydrogen, known as the Radio Recombination Lines (RRL's), can make it possible to study HII regions that would otherwise be obscured by the intervening interstellar medium, which generates so much extinction to optical telescopes. When observing RRLs several assumptions can be made:

- RRLs are weak and can be observed at the optically thin limit.
- The downward transition is a strong function of quantum number, and the time scale for de-excitation becomes very long with high quantum numbers.
- At radio wavelengths the Rayleigh–Jeans approximation to the Planck function can be used.

The optically thin emission of the source holds information about the temperature and density of the gas cloud, but the observation of a single RRL is unable to provide enough information to be able to determine one property without knowing the other first. However, the same ionized gas will emit thermal Bremsstrahlung continuum emission. The ratio of line intensity to continuum emission in the vicinity of the line can yield a value for the electron temperature.

The HI 21 cm Emission Line of Hydrogen

The quantum energy change involved in the emission of 21 cm radiation is known as the hyperfine transition of the electron ground state, where the electron spontaneously reverses its spin property. The electron ground state is where the electron is at its closest to the proton nucleus. Therefore it originates from the cold hydrogen that pervades the interstellar medium. This cold hydrogen exists as single hydrogen atoms, unlike the molecular H_2 involved in the ionized clouds surrounding young stars.

The classical model of the hydrogen atom is that of an electron orbiting a proton; the electron is spinning either clockwise or anticlockwise. In the quantum age we now know this model is not accurate, but it is still a useful way of picturing the process in the mind's eye. The hydrogen atom has slightly higher energy if the spin of the electron and proton are the same (parallel spin), and slightly lower energy when the spins are anti-parallel. As with many of the hyperfine transitions emanating from a split quantum state, it is known as a forbidden line. As such the time interval between transitions is extremely long, in the order of 10 million years or more in this case. However, hydrogen is so common in the universe that the radiation is easily observed by radio telescopes.

The rest frequency of emission is 1420.40575177 MHz in the L band. However the Doppler shift in frequency is easily observed, which can quickly give a measurement to the rotation speed of our Milky Way. Theoretically, at least, the study of red-shifted 21 cm emission in the range of 200 MHz down to the ionospheric cutoff about 10 MHz can be used to probe the state of the early universe. In practice this is extremely difficult from Earth's surface due to interference from manmade sources and in isolating the weak signals from the background continuum previously discussed.

Following is a table of the most important molecular emission lines in the radio spectrum up to a maximum of 100 GHz.

Molecule	Rest frequency
Deuterium (DI)	327.348 MHz
Hydrogen (HI)	1,420.406 MHz
Hydroxyl radical (OH)	1,612.231 MHz
Hydroxyl radical (OH)	1,665.402 MHz
Hydroxyl radical (OH)	1,667.359 MHz
Hydroxyl radical (OH)	1,720.530 MHz
Methyladyne (CH)	3,263.794 MHz
Methyladyne (CH)	3,335.481 MHz
Methyladyne (CH)	3,349.193 MHz
Formaldehyde (H_2CO)	4,829.660 MHz
Methanol (CH_3OH)	6,668.518 MHz
Helium ($^3He^+$)	8,665.650 MHz

(continued)

Molecule	Rest frequency
Methanol (CH_3OH)	12.178 GHz
Formaldehyde (H_2CO)	14.488 GHz
Cyclopropenylidene (C_3H_2)	18.343 GHz
Cyclopropenylidene (C_3H_2)	21.587 GHz
Water vapour (H_2O)	22.235 GHz
Dicarbon monosulphide (CCS)	22.344 GHz
Ammonia (NH_3)	23.694 GHz
Ammonia (NH_3)	23.723 GHz
Ammonia (NH_3)	23.870 GHz
Ammonia (NH_3)	24.139 GHz
Methanol (CH_3OH)	36.169 GHz
Cyanoacetylene (HC_3N)	36.392 GHz
Silicon monoxide (SiO)	42.519 GHz
Silicon monoxide (SiO)	42.821 GHz
Silicon monoxide (SiO)	42.880 GHz
Silicon monoxide (SiO)	43.122 GHz
Silicon monoxide (SiO)	43.424 GHz
Dicarbon monosulphide (CCS)	45.379 GHz
Cyanoacetylene (HC_3N)	45.490 GHz
Carbon monosulphide (^{13}CS)	46.247 GHz
Carbon monosulphide ($C^{34}S$)	48.207 GHz
Carbon monosulphide (CS)	48.991 GHz
Oxygen (O_2)	56.265 GHz
Oxygen (O_2)	58.324 GHz
Oxygen (O_2)	58.446 GHz
Oxygen (O_2)	59.164 GHz
Oxygen (O_2)	59.591 GHz
Oxygen (O_2)	60.306 GHz
Oxygen (O_2)	60.434 GHz
Oxygen (O_2)	61.151 GHz
Oxygen (O_2)	62.486 GHz
Deuterated formylium (DCO^+)	72.039 GHz
Deuterium cyanide (DCN)	72.415 GHz
Cyanoacetylene (HC_3N)	72.784 GHz
Methyl cyanide (CH_3CN)	73.59 GHz
Deuterated water (HDO)	80.578 GHz
Cyanoacetylene (HC_3N)	81.881 GHz
Cyclopropenylidene (C_3H_2)	82.966 GHz
Cyclopropenylidene (C_3H_2)	85.339 GHz
Methyl acetylene (CH_3CCH)	85.5 GHz
Deuterated ammonia (NH_2D)	85.926 GHz
Hydrogen cyanide ($HC^{15}N$)	86.055 GHz
Silicon monoxide (SiO)	86.243 GHz
Hydrogen cyanide ($H^{13}CN$)	86.340 GHz
Formylium ($H^{13}CO^+$)	86.754 GHz
Hydrogen isocyanide ($HN^{13}C$)	87.091 GHz
Silicon monoxide (SiO)	86.847 GHz
Ethynyl radical (C_2H)	87.300 GHz
Hydrogen cyanide (HCN)	88.632 GHz
Hydrogen isocyanide($H^{15}NC$)	88.866 GHz
Formylium (HCO^+)	89.189 GHz
Hydrogen isocyanide (HNC)	90.664 GHz
Cyanoacetylene (HC_3N)	90.979 GHz
Methyl cyanide (CH_3CN)	91.98 GHz

(continued)

Molecule	Rest frequency
Carbon monosulphide (^{13}CS)	92.494 GHz
Diazenylium (N_2H^+)	93.174 GHz
Carbon monosulphide ($C^{34}S$)	96.413 GHz
Carbon monosulphide (CS)	97.981 GHz
Sulphur monoxide (SO)	99.300 GHz

The 3 K Microwave Background

Our Milky Way Galaxy is pretty noisy in the HF and VHF. The sky background brightness when expressed as a temperature is hundreds of kelvin at a wavelength of 1 m, but at a wavelength of 0.1 m the cold sky background is only 3 K. This signature is the background radiation left over from the Big Bang, which formed the universe. This temperature, therefore, establishes a limit to the sensitivity that is possible for a radio telescope. In fact in most cases the noise performance of a radio telescope will be limited by the thermal noise produced within the first amplifier stage for a telescope operating at centimeter wavelengths or less. Professional instruments are cooled with liquid nitrogen to improve their performance.

Radio telescopes working at meter wavelengths can cope with much higher noise levels generated within the electronics, but ideally the first amplifier stage (which contributes most to the overall noise performance) should be better than the background sky level. In practice this should not be difficult to achieve.

Pulsars

Most of this chapter so far has covered general topics on the observation of the Milky Way and extragalactic objects, the bread and butter radio sources being ionized gas clouds or neutral cold gas clouds. One type of object warrants special attention – the pulsars.

The first pulsar was discovered in 1967 by Jocelyn Bell and Anthony Hewish using a phased array of 2,048 dipole antennae spread over a 4-acre field. They were working at a frequency of 81.5 MHz. Pulsars are cataloged with a prefix of PSR, followed by their location in right ascension and declination. For example the first pulsar discovered by Jocelyn Bell at Cambridge is PSR 1919+21. This decodes to a right ascension of 19h19m and a declination of +21°.

Pulsars show the following properties:

- Most have periods between pulses of between 0.25 and 2 s, although some millisecond pulsars are known, and the longest is over 8 s.
- The pulse period is extremely stable and repetitive.
- Pulsars spin down over time but very slowly. Their characteristic lifetimes (the time it takes for pulses to stop if the spin down rate remains constant) varies but are expected to be hundreds of millions of years.

Pulsars were discovered accidentally during an experiment to investigate scintillation of radio waves. The almost artificial regularity of the pulse train suggested alien intelligence, although this was quite quickly ruled out. The very rapid pulses could only be explained by a very compact spinning object. This ruled out the possibility of their origin from binary star motion, pulsating variable stars, or even conventional stellar rotation rates. Thomas Gold postulated the emission was generated from a rapidly rotating neutron star.

Indeed, within a year of discovery, pulsars were detected within the Vela and Crab nebulae supernovae remnants. Once an average-mass star runs out of fuel, the core collapses, and the outer gaseous envelope is blown off in either a planetary nebula or by a supernova, depending upon the initial mass. The collapsed core of objects less than 1.4 solar masses form white dwarves – consisting of a degenerate compact soup of free protons, neutrons, and electrons. However, if the mass is between 1.4 and 3 times solar masses, the gravitational contraction is sufficient to increase the density so much as to force the nucleons very close together. This converts all protons into neutrons. The result is a degenerate soup of neutrons, quarks, etc., and is known as a neutron star. The neutron "fluid" is expected to be superconducting at temperatures up to 10^9 K. More massive stars after a supernova explosion would form black holes. A neutron star of 1.4 solar masses with a period of 1.4 ms could be as small as 40 km across, though the average object would be somewhat larger. They are all considerably smaller than Earth.

How do we know this? After all, neutron stars have not been observed directly.

Consider the rotational period of the neutron star. The pulse period gives a direct measurement of the rotational period of the star. From simple physics, the minimum radius would be such that the centrifugal forces would balance the gravitational contraction, or else it would break up. The figures calculated for a 1.4 solar masses object reveal radii as small as 20 km. Now that the minimum dimension is known the minimum density can be estimated. From this, the period of the first pulsar to be discovered (1.3 s) would suggest a density consistent with white dwarf stars. However, once faster objects such as the Crab Nebula ($p = 0.033$ s) were found, the implied density of $>10^{14}$ g cm^{-3} could not be explained as a white dwarf and so must be a neutron star. From the observations of several pulsars whose mass could be determined (from their mutual revolution around other massive objects nearby) they all appeared to be very close to 1.4 solar masses. These high spin rates are accounted for by the conservation of angular momentum as the core collapses.

The mechanism of radio generation is poorly understood but will involve synchrotron processes. You will be familiar with Earth and its magnetic field. The rotational axis of Earth does not match that of the magnetic field. We have already seen for Jupiter, the planetary dipole field is tilted about 10° with respect to Jupiter's rotational axis; this situation is common and is true of neutron stars and therefore pulsars, too. Energy is beamed from the magnetic poles, in a north and south direction. As the neutron star rotates, the radio beam sweeps quickly past our line of sight, and we see it as a pulse of radiation. If we assume that all neutron stars beam radiation from their poles, clearly then we cannot observe all neutron stars by means of pulsed radio output, but statistically a few will have a suitable orientation for us to study them.

Pulsed output from neutron stars is not restricted to radio energy but can be seen throughout the electromagnetic spectrum, at least in some cases. The Crab pulsar can even be seen flashing in the optical spectrum, although very high time resolution is required, or the use of a rotating shutter to effectively slow down the pulses. The white light glow of the surrounding nebula shows significant polarization and is also associated with synchrotron emission, which is consistent with a magnetic field within the cloud of around 10^{-3} G.

This was puzzling at first, because the expansion of the nebula should have considerably weakened the field by now. The amount of power needed to maintain the expansion, the relativistic electrons, and the magnetic field is many times the output of our own Sun. All this power is coming from the pulsar. The pulsed radiation from the poles could not be the primary mechanism powering the nebula because it is 200 million times too small. It turns out the amount of power required to explain the properties of the nebula balance with the amount of power lost from the neutron star due to its spin down rate.

One idea for the energy loss mechanism is that the rotation of its magnetic field (polarization of the radio waves shows there is a field) induces a complementary electric field at a distance from the star. This forms an electromagnetic wave called magnetic dipole radiation. The magnetic field of a neutron star is expected to be very great, due to the conservation of magnetic flux, as the core initially collapses. This can magnify the strength of the field by as much as 10^{10} times. It is possible that a dynamo-like effect may increase this even more. The magnetic dipole radiation generated from the strong field would be of very low frequency, <1 kHz. Not only would that not penetrate Earth's atmosphere, it would not even escape through the interstellar medium. However, it is thought that magnetic dipole radiation is the primary energy loss mechanism driving the slow spin down rates of pulsars.

Certainly there is a substantial magnetosphere around the pulsar. Huge electric forces pull electrons and charged particles from the star. The magnetosphere co-rotates with the magnetic field, and the angular velocity will therefore increase with distance from the neutron star. However, the velocity can never exceed the velocity of light, so particles in the outer magnetosphere are spun away in a sort of wind, carrying the magnetic field with them into the surrounding cloud.

The Crab pulsar is relatively young; in fact, the supernovae was noted in historical records in 1054. These young pulsars show random glitches in the rotation period, where the periods decrease by between 10^{-6} and 10^{-8} s, with intervals of several years. The pulses are sharp and short, accounting for only 1–5% of the pulse period in time. However interaction with electrons in the interstellar medium slows down the radio waves, rather like light slows down when passing through a higher density refractive medium. The slowing process is wavelength dependent. Longer wavelengths slow down most. This turns a sharp "tick" of a pulse into a drawn-out descending pitch whistle. The amount of dispersion can be used to estimate the pulsar's distance. Pulse shapes vary widely, in that individual pulses can be a group of close sub-pulses. The sub-pulse structure may also vary with time, although by averaging these pulses over 100 or more cycles, the overall shape remains stable.

Pulsars are exceedingly difficult to detect. Their initial discovery was a very lucky accident, noted by a very keen-eyed astronomer. Jocelyn described the first recordings as "a bit of scruff," a short burst of energy recorded on a pen chart where she noted its regularity, and its return 23 h and 56 min later guaranteed its celestial origin. In the major radio surveys conducted by the large radio telescopes the typical limiting sensitivity was about one Jansky. By a sheer coincidence, the mean flux density at meter wavelengths of the strongest pulsars is about one Jansky. Let's say the background electrical noise in the 250 foot mark 1 Jodrell Bank telescope was 100 Jansky, which itself is operating with a noise temperature of 100 K. Pulsars therefore have a signal 10^{-4} times that of the noise, and even then the pulses are only near their peak output up to 5% of the time. If the bandwidth was 1 MHz and the integration time of the receiver was, say, 10 ms, which is reasonable, then the sensitivity would be 100 times better than the noise, but in order to observe a single pulse it would still have to be considerably stronger than one Jansky and lasting more than 10 ms. Clearly special techniques were required.

Increasing the integration time did not help. In order to observe detail in the pulses more than two samples would be required in a pulse period, ideally considerably more. One technique that was used was to integrate the signal by superposing pulses together over a period of time. It helps to already know the period, of course, but with heavy post processing and trial and error the period of an unknown pulsar could be obtained.

Although observing pulsars at meter wavelengths is exceedingly difficult, the problem eases somewhat in the microwave region. Pulsars can be observed by radio methods at frequencies of between 20 MHz and 10 GHz. At least in the microwave region the sky is very cold and noise free. The telescope is then limited only by the local electrical noise and the noise temperature of the telescope front end. Low noise amplifiers and cooling with liquid nitrogen makes the job of pulsar studies easier.

Chapter 5

Antennae

The antenna is the element that collects radio energy from its surroundings and converts it into an oscillating electrical signal in the tube or wire from which it is constructed. It is one of the most critical pieces of technology for any radio receiver. Get the antenna wrong, and no matter how good the receiver is, it will not perform at its best.

Antennae appear to be simple constructions of wire or tube, but proper understanding of their function and design is amazingly complex at times. There are many texts available on the subject, which could on its own fill this entire book. Presented here are several common designs that can be adapted easily to the frequency of choice.

All antennae have the following properties:

- Characteristic impedance
- Directional properties
- Forward gain
- Polarization
- Beam pattern

Characteristic impedance is described in ohms. The concept of impedance is discussed in more detail in the electronics chapter. Impedance of the antennae should be closely matched to that of the feeder cable, which in turn should be closely matched to the impedance of the input side of the receiver. Mismatched impedances at any point create internal reflections of the signal, reducing the amount of energy passed on to the next stage and thereby reducing the efficiency of the system.

The actual impedance of a practical antenna is complex and can vary with its proximity to the ground or other nearby structures. To aid in setting up you can purchase antenna analyzers from amateur radio supply stores that will allow the constructor to adjust and match systems for best efficiency. However these analyzers are set up mostly for amateur radio bands. Later we will describe the construction of an impedance bridge you can build yourself.

Although all antennae have some directional properties, it is desirable in radio astronomy to concentrate on systems that have significant forward gain. The directional properties and forward gain are closely related. So called omnidirectional antennae, often used by amateur radio enthusiasts for scanning the airwaves or for general communication, strictly speaking, do not radiate equally in all directions. Such antennae are often vertical and do not radiate well straight up, but then no

J. Lashley, *The Radio Sky and How to Observe It*, Astronomers' Observing Guides,
DOI 10.1007/978-1-4419-0883-4_5, © Springer Science+Business Media, LLC 2010

one is transmitting above you, are they? Except maybe from outer space! For this reason we will not be discussing vertical or omnidirectional antennae here. Forward gain is a numerical measure of how much more efficient an antennae is in a beamed direction than an ideal omnidirectional (or isotropic) unit would be. The value is expressed as either a pure number (a ratio) or more often in the decibel notation dBi, where the "i" refers to the ideal isotropic radiator.

It is important to remember that all real antennae are polarized. To understand the meaning of this, consider the nature of the electromagnetic (EM) waves. An EM wave travels in a straight line in free space (for our purpose Earth's atmosphere can be considered free space, too); perpendicular to the path is an electric wave, and perpendicular to both is a magnetic wave. For the most part the antennae we use on conventional radios couple to the electric wave component of the radio signal. Radio signals are linearly polarized if all the electric wave components are parallel. For proper reception a linearly polarized antenna, such as a wire dipole, should be aligned with the electric vector of the impinging waves. In practice, for radio astronomy signals are often randomly polarized. This means that a single antenna will only collect half of the potential radiation.

The beam pattern of an antenna is another complex story. Beam pattern for, say, a simple dipole can be altered by changing its length! This pattern can become narrower and more complex by adding several antennae in an array close to each other. The beamwidth of a practical antenna is expressed in degrees. It is the similar to field of view of an optical telescope. Beamwidth is an angular measure between the half power points (3 dB points). The half power points are those "off axis" points at which the received power drops to half of the value of the center of the pattern. Note here that not all directional antennae have circular beam patterns.

When designing or evaluating an antenna system for a radio telescope – which is, of course, a radio receiver – it is often more useful to think of the antenna as if it was transmitting instead. For example, the majority of antennae have some directional properties called the beam pattern. It is easier to picture the beam pattern as if it were radiating away. The same beam pattern works in reverse for the receiver.

The Dipole

Although the simplest antenna is a long, straight wire, which will certainly pick up radio to some extent, it will not be considered here. A more practical antenna is the dipole, consisting of two lengths of wire or tube (later we will see a folded version) whose center connections are linked to the radio transmitter or receiver.

The most common dipole antennae are half a wavelength wide for the wavelength of interest. For example, let's take a Jovian decametric receiver working at 20.5 MHz. The length of a dipole can be easily calculated by this formula

$$L = \frac{143}{f}$$

where f is the frequency in megahertz and L is the total length of the dipole in meters.

This gives a dipole length of 6.98 m. Note that the formula takes into account some foreshortening effects of real antennae, so the resulting length is slightly shorter than half a wavelength.

The radiation pattern of a dipole is, when looked at in two dimensions, that of a figure of eight, with "lobes" extending perpendicular to the antenna. In three dimensions the radiation pattern is more like a doughnut. The key fact here is that a dipole does not work well in a direction parallel to itself. Equally, any noise within the receiving bandwidth which is in line with the dipole will be significantly suppressed. Because the antenna receives signals in a preferred direction and suppresses signals in other directions, and the length of it is tuned to the wavelength of interest and is less efficient at other wavelengths, we say it has gain. In our example, the dipole, the gain will be 1.64. Gain is a pure number, a ratio of the radiated power (remember it is easy to think in terms of a transmitting antenna) divided by the radiated power of an ideal isotropic antenna (one that radiates equally in all directions). Hence our dipole will provide a signal strength of 1.64 times better, in its beam direction, than a unit that is omnidirectional. The more directional the antenna is, the greater is its gain. This gain is free – it does not rely on electrical power!

You might be thinking, why half a wavelength? Why not a full wavelength? Clearly for the dipole to be tuned to the wavelength of interest, it must have a length that is a clean fraction of receiving wavelength, such and a quarter, a half, a three-quarter, etc. Well, the effect of dipole length is not what you may be expecting. A quarter wavelength dipole has a wider beam pattern than a half wavelength version, and so it has less gain, but they both have that simple doughnut radiating pattern. A full wavelength dipole actually has a double doughnut radiating pattern and is most sensitive at a 45° angle! The three-quarter wavelength version is a hybrid, with weaker perpendicular lobes and stronger 45° lobes (see Fig. 5.1). Hence the half wave version gives us a nice simple radiating pattern.

A note of caution, then, for our example dipole working at 20.5 MHz. The 6.98-m dipole presents a full wavelength dipole at 41 MHz and will therefore be sensitive at 45° intervals at this higher frequency. The Jovian receiver should be suitably filtered at its input with at least a low-pass filter with a cutoff point below 41 MHz, and better still with a bandpass filter having a low cutoff above 10.25 MHz and a high cutoff below 41 MHz.

It is important to understand how to connect the dipole antenna to the receiver. All modern receivers have what is known as an unbalanced input. In plain words, this means that one of the input conductors is grounded to Earth (the outer conductor of a coaxial connector). The dipole, on the other hand, is balanced – neither half is grounded. If we were to just connect our coaxial feed cable to the center contacts of the dipole, the feeder would become part of the antenna and in the transmitting case would radiate energy. It would also affect the beam pattern for a receiving antenna and alter its tuned length. To combat this problem we need to provide a balun transformer.

The word balun comes from "balanced to unbalanced." In the case of the dipole antenna considered the center point impedance is about 75 Ω (although this will vary with height above the ground). This matches quite well to 75 Ω coaxial cable. The balun required is therefore a 1:1 balun, meaning it does not change the characteristic impedance. As we will see later a folded dipole will show an impedance of 300 Ω, so a 4:1 balun is used to transform the impedance down to 75 Ω. The properties of a balun provide equal but opposite phase currents to flow in the feed cable; therefore, the effects cancel each other out, and the feeder does not radiate power or affect the tuned length of the dipole.

Baluns can easily be purchased from amateur radio supply stores, but are equally easy to make. A simple ½ wave loop of coaxial cable, the same used to feed the receiver, can be used. The diagram below shows how this is done (Fig. 5.2).

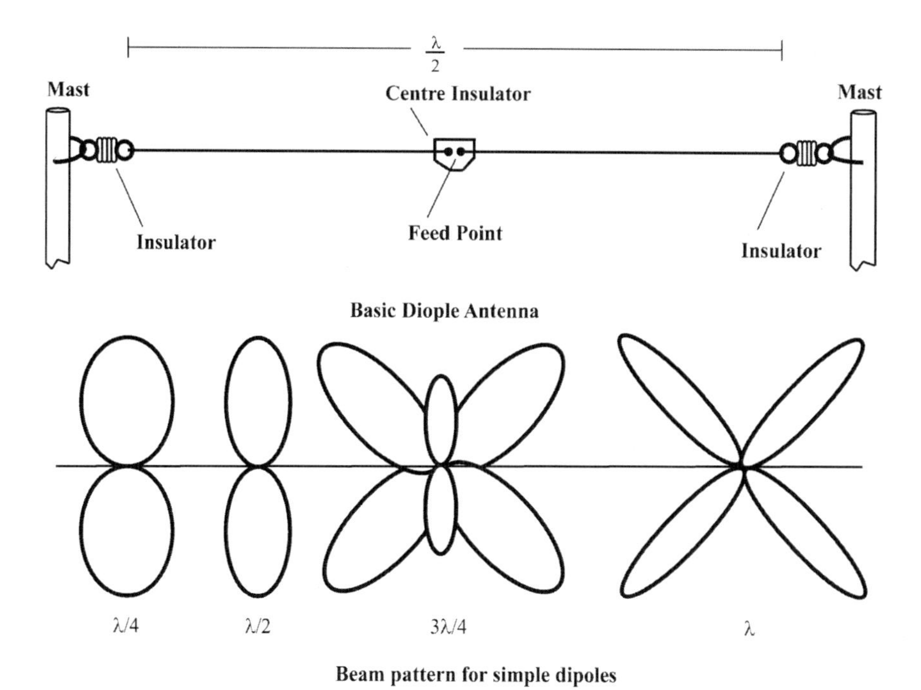

Fig. 5.1. Basic wire dipole with it radiation patterns. Usually the dipole is cut to a half wavelength, which provides the simplest beam pattern with reasonable gain.

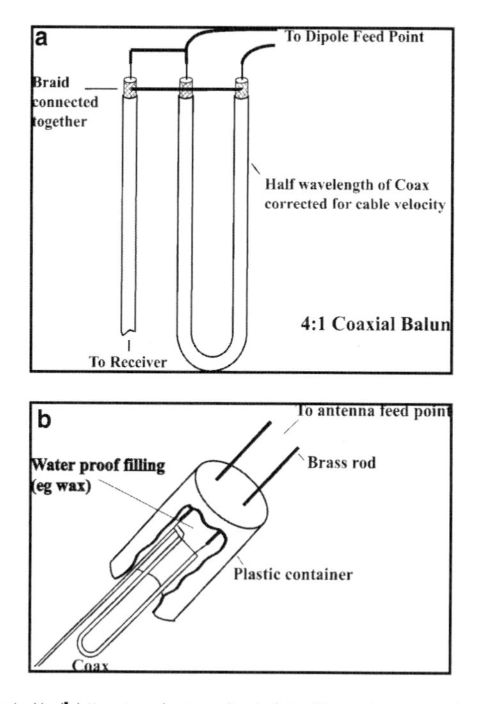

Fig. 5.2. (**a**) Balun coaxial cable. (**b**) Mounting and waterproofing the balun. If using silicone as a sealant build it up in layers to ensure it dries properly.

A Simple Dipole

The simple dipole is a resonant antenna with a very small bandwidth. It is tuned to one wavelength. This is fine for a fixed channel receiver dedicated to one job, but certain applications may require the receiver to be tunable over a range, to cover, say, a band of wavelengths allocated to radio astronomy, or so the spectrum either side of a center frequency can be studied. Going back to our Jovian decametric receiver scenario, let's look at how we can improve the dipole. In order to avoid local interference at 20.5 MHz we may want to tune the receiver between 18 and 24 MHz to find a suitably quiet spot away from ground-based channels. We can make three dipoles in one go with a common feed point. So let's make one for 18 MHz, one for 21 MHz, and the third for 24 MHz. The resulting lengths are then 7.94, 6.81, and 5.96 m. This technique works so long as any one dipole is not an odd integer multiple length of any of the others. A clever way to build this antenna is to use three core cable cut to the longest dipole, then shorten the other two cables to suit. At the center the three cores are connected together on each side.

A Folded Dipole

An alternative to multiple dipoles would be a single folded dipole. This is illustrated in the Fig. 5.3. Folding has two advantages; firstly, it is a much wider bandwidth, and secondly, it is narrower than the single equivalent.

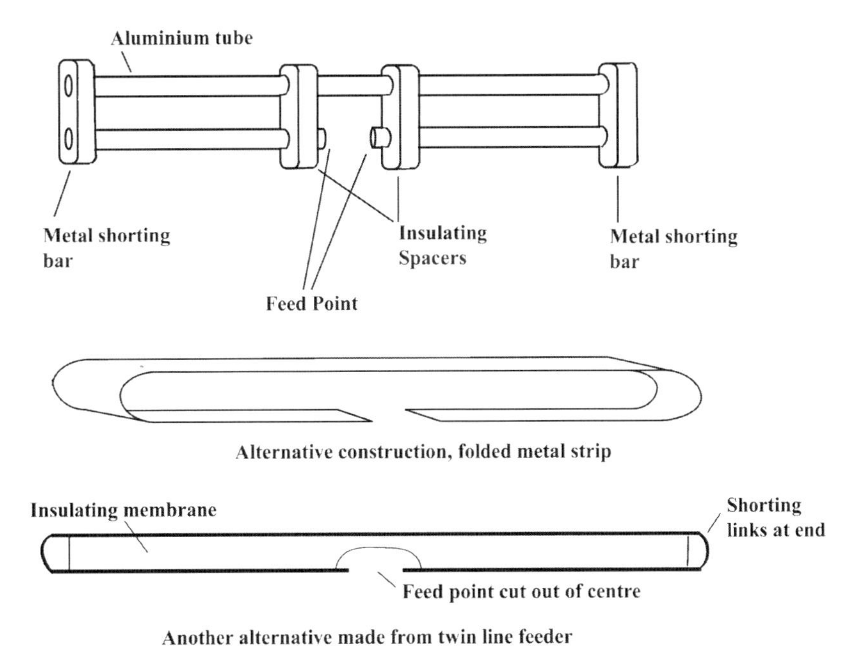

Fig. 5.3. Folded dipole ideas. The aluminum tube version is suitable for short wavelength because of its high rigidity. The twin line version suits long wavelengths, because it is lightweight. The twin line is automatically mounted parallel by virtue of its construction around the plastic sheet insulator. (For increased bandwidth extra shorting bars or wires can be added; see text for details).

The characteristic feed point impedance is now 300 Ω, so the balun must now convert the impedance as well. A 4:1 balun divides the impedance of the antenna by four and therefore matches well with 75 Ω coaxial feeders such as RG59. It can be constructed using a 300 Ω twin feeder, which used to be used for television applications. A length is cut to the total dipole dimension, the ends are shorted at both sides, and one of the conductors is broken at the center point, which becomes the feed point. Most 300 Ω cable is quite weak in construction and when subjected to the weather and wind loading forces may break if suspended on its own. It is advisable to provide support at least at the center and the tips. The diagram illustrates some techniques of construction and mounting. The total length of the loop should be as before, but the linear length is about half that of the unfolded version. A trick to increase the folded dipole bandwidth even more is to add an extra pair of shorts closer to the feed point. The distance from the feed point to these inboard shorts can be calculated from

$$L = \frac{61}{f}$$

where L is the length in meters, and f is the frequency in megahertz.

Folded dipoles are most often seen in the construction of Yagi antennae, particularly for UHF television reception. The improvement in bandwidth is suited to cover the range of UHF television channels employed in a given region. Yagis will be discussed in more detail later.

It is possible to improve the performance of a dipole and increase its forward gain and narrowing its directionality by adding a corner reflector. In theory a reflector would be a parabolic cylinder with the dipole at its focus. However, at the long wavelengths we are dealing with it turns out that a simple 90° corner reflector is a good enough approximation and works very well. The reflector should be slightly wider than the dipole, and the depth of the sides should not be less than 0.7 wavelengths. The dipole is placed at the focus of the corner, which is calculated as 0.35 wavelengths from the apex of the corner. The frame of the reflector can be constructed from wood. It needs to be lined with wire mesh or parallel wire lines. Wire line could be low-cost galvanized steel fencing wire and should be parallel to the dipole. The separation of the wires or the size of the mesh should be less than or equal to 1/10 of a wavelength (Fig. 5.4).

Inverted V Dipole

Normally wire dipoles are mounted horizontally, but they can be mounted in the form of an inverted V shape. A single mast is used to support the feed point, and the dipole "arms" are drooped at an angle of between 90 and 120° towards the ground. As with all such antennae, tip insulators (sometimes called dog bones) should be used. The best quality insulators are porcelain, although plastic ones can be found. Ropes attached to the insulators are then staked into the ground securely. The radiating pattern of the inverted V is essentially the same, but owing to the bend it is spread over a wider area. If the Jovian decametric receiving antenna were mounted this way on an east-west base line, it would increase the duration in

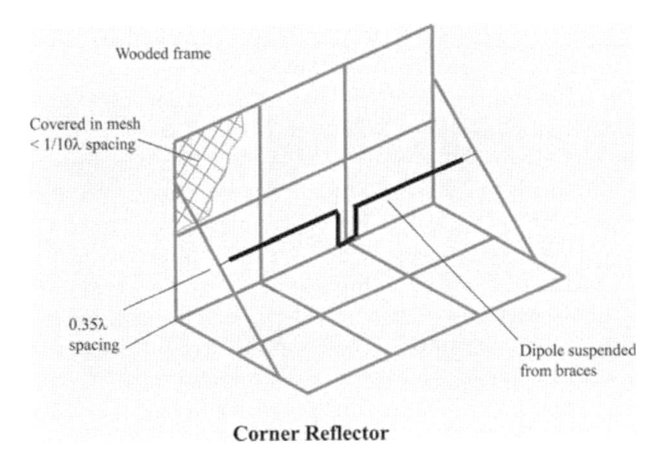

Fig. 5.4. Dipole corner reflector. Wire mesh needs to be finer pitch than 1/10 wavelength. The dipole is suspended 0.35 wavelengths above the center between both sides. Use a diagonal cross brace on each side to support it.

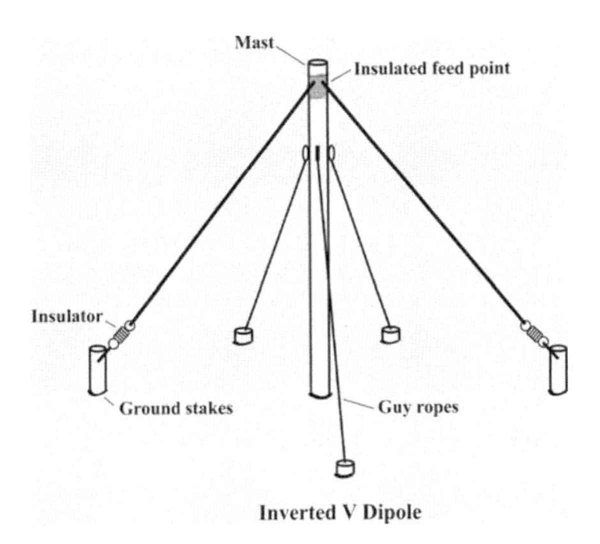

Fig. 5.5. Inverted V dipole.

which Jupiter would pass through the stronger part of the beam over an equivalent horizontal version. In this case the dipole length should be the full half wavelength long calculated from L = 150/f (Fig. 5.5).

There are many other variations on the dipole antenna out there. A common one is known as a G5RV. However, most of these more complex units are used by amateur radio enthusiasts to provide access to multiple frequency bands from a single antenna. Commercially made units will be tuned to amateur frequency bands and therefore would be less suitable for use in radio astronomy; they also require variable antenna tuners. In radio astronomy observing it is preferable to build an antenna tuned for one task, and dedicate it to that one job. The simple dipoles presented here are relatively narrowband and therefore have the advantage of rejecting a lot of unwanted noise. However, always keep in mind that these simple dipoles are resonant on odd harmonics of the design frequency, too. If one of these harmonics falls in a particularly noisy RF band it could cause undue interference. Therefore the receiver should be supplied via an appropriate low pass filter. Filter design is dealt with in a later chapter.

The amount of signal collected by an antenna is often expressed as its collecting area in square meters. It is not at first obvious how this relates to a dipole antenna. The definition of collecting area is the ratio of received power divided by the intensity of the collected wave. For a dipole this translates to about $0.13\lambda^2$, so for a working frequency of 100 MHz and a wavelength of 3 m the collecting area would be about 1.2 m^2. For a working wavelength of 1 m this would reduce to 0.13 m^2. Not very much considering the weak celestial sources we deal with.

One way to improve the situation is to increase the number of dipoles used and create an array of them. This quickly begins to demand a lot of space, though, because they need to be adequately separated from each other to avoid undue interaction playing all sorts of tricks. For one thing, if the aerials are too close together, the feed point impedance is badly affected, making impedance matching difficult. If they are spaced at least one wavelength apart in a horizontal line, then their combined effect is to multiply the collecting area by the number of dipoles used. Note, however, that the result is strongly sensitive along the line of the array. Not a great idea, as it is a potential source of interference from terrestrial sources. If the inter-dipole separation is now reduced to 0.8 wavelengths, the horizontal sensitivity is much reduced, but the central lobe expands from about 14° to a little over 17°. The beam pattern also becomes asymmetrical, as it is narrower in the horizontal plane and wider in the vertical plane. Array antennae are complex devices and a difficult area for the novice constructor. It takes a lot of care to ensure the interconnections are phase-matched correctly. More detail can be found in the ARRL antenna handbook and similar texts.

Large Loop Antennae

Loop antennae are often built in a square form, which lends itself to its common name of the Quad, or Quad Loop. A small multi-turn loop is discussed in the VLF receiver chapter. The larger cousin can be used at HF or VHF frequencies quite efficiently. Here the length of the loop from feed terminal to feed terminal is one full wavelength. The gain of the Quad is about 2 dB more than the half wave dipole. The radiating pattern is a twin-lobe form projecting from the faces of the loop. The total length of the loop can be calculated from the following formula:

$$L_A = \frac{306.4}{f}$$

where L_A is the length in meters and f is the frequency in megahertz.

Note here that these large loops use the electric field of the radio waves to function. In the case of the small VLF loop the coupling is magnetic and is therefore directional in the plane of the loop (Fig. 5.6).

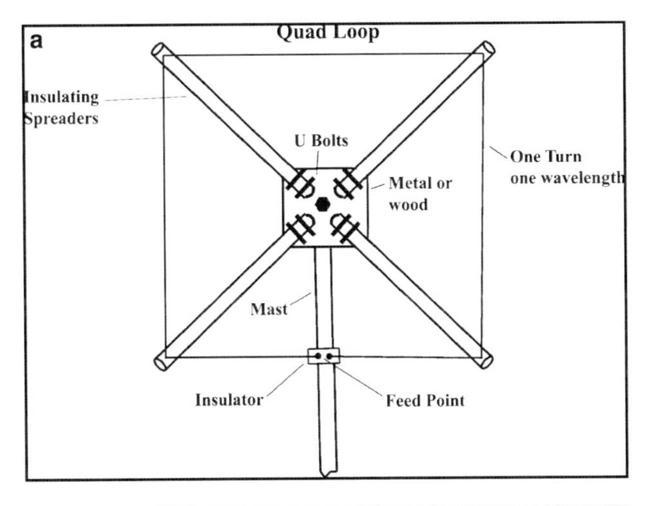

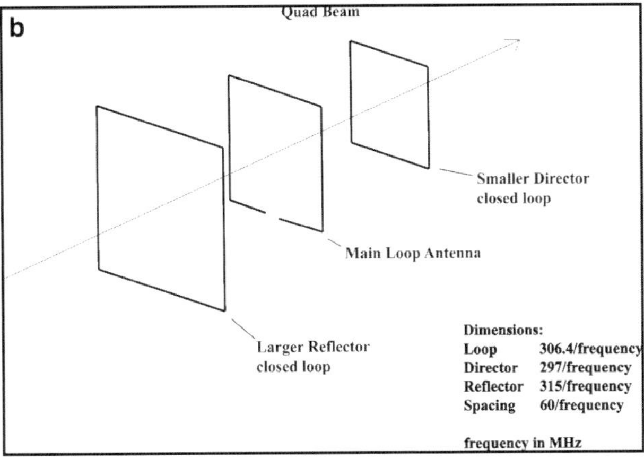

Fig. 5.6. (**a**, **b**) Large Quad loop antenna. By adding a reflector and a director the forward gain can be increased.

The loop could be constructed using a cross made from insulating material such as wooden dowels or marine plywood (weatherproof) strengthened at the center by a pair of square gusset plates. The Fig. 5.6 illustrates the details. The central gusset plates make it possible to mount it onto a mast.

The Quad is bidirectional. It is therefore advantageous for us to consider ways of making it unidirectional and more efficient in that direction at the same time. For the most part in radio astronomy, the objects we wish to observe are located in well defined directions, which move in unison with Earth's rotation. The more directional our antennae are, the better we are able to pull in the weak signals we seek.

For the Quad loop a reflector loop can be placed behind the main loop. The reflector is slightly larger than the main (or driven) loop. Its length can be calculated from the following formula:

$$L_R = \frac{315}{f}$$

In addition and less intuitively, the addition of a director loop in front of the main one further improves the forward gain. The length of this loop is given by:

$$L_D = \frac{297}{f}$$

Once again the units of L are in meters and f in megahertz.

The spacing between the three elements in meters is given by:

$$S = \frac{60}{f}$$

The resulting Quad beam antenna will have a forward gain of about 9 dBi. Some say the Quad beam works better than the Yagi antenna when it is close to the ground (less than half wavelength). The feed point impedance will be around 60 Ω, so this will be a fair match to either 75 or 50 Ω feeders, although a 1:1 balun will still be required.

Note here the Quad will be horizontally polarized if the feed point is at the center of the bottom section or the center of the top section; it will be vertically polarized if fed from either side.

Yagi Beam Antenna

The classic example of a Yagi antenna, or to give it its full name of the Yagi-Uda Array, is the UHF television aerial. Indeed, if you were planning to work in this range of frequencies it would make sense to buy a high-gain antenna off the shelf. This section discusses the design of Yagis in general so that you can experiment at other frequencies.

The Yagi is characterized by the number of elements, the elements being the dipole, reflector, and any directors used. The minimum is two elements, consisting of the driven dipole and a passive reflector behind it, or a dipole with a single director. Consider for now a two-element system with a straight unfolded dipole. Placing a reflector 5% longer behind with a spacing of 0.2 wavelengths from the dipole would yield a peak gain of a little less than 5 dBd (5 dB better than the dipole alone). The gain variation is fairly insensitive to spacing. Placing only a director (5% smaller) in front by 0.1 wavelengths would provide a gain of a little over 5 dBd. The gain this time would be quite sensitive to spacing. However the front to back ratio of a two-element design is quite poor. The feed point impedance is also a complex function of element spacing. Yagi design is far from simple; even after running calculations real world antennae can often be improved by spacing adjustments carried out by practical experiment.

Yagi's are most often used at VHF or higher frequencies, where their size is relatively compact. Here are the formulae for calculating the element dimensions and spacing for a two-element system.

$$Director_{2e} = \frac{138.6}{f}$$

$$Dipole_{2e} = \frac{146}{f}$$

$$Spacing_{2e} = \frac{44.98}{f}$$

where f is the frequency in megahertz and the dimensions provided are in meters.

Combining a reflector and director into a three-element beam should improve the front to back ratio and provide about an 8 dBd gain. The dimensions now are given by the following formulae:

$$Director_{3e} = \frac{140.7}{f}$$

$$Dipole_{3e} = \frac{145.7}{f}$$

$$Reflector_{3e} = \frac{150}{f}$$

$$Spacing_{3e} = \frac{43.29}{f}$$

where again dimensions are in meters and f is in megahertz.

The impedance at the feed point will be between 18 and 26 Ω, so some form of matching will be required to a coaxial feeder.

An extra 1 dBd of gain may be achieved by adding a second director, but the front to back ratio will be poorer unless the element spacing is increased; adding three directors oddly does not improve the situation much over a four-element, but a six-element has good gain and a good front to back ratio at the expense of extra length, making this impractical in the HF band.

The calculations for a six-element Yagi are:

$$Director_{6e} = \frac{134.39}{f}$$

$$Dipole_{6e} = \frac{145.05}{f}$$

$$Reflector_{6e} = \frac{145.56}{f}$$

$$Spacing_{6e} = \frac{44.8}{f}$$

where the dimensions are again in meters and the frequency f in megahertz.

The collecting area or efficiency of the Yagi is much improved over the open dipole and approximates to $0.65\lambda^2$ (Fig. 5.7).

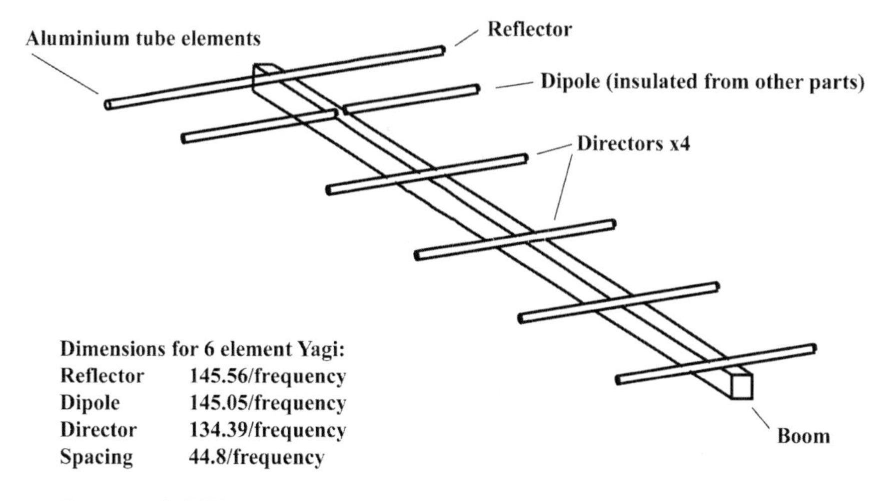

Fig. 5.7. Basic six-element Yagi beam. For the basic Yagi the dipole is split inside the boom where the feed point is, so it may need additional external support; however, if a gamma match is used the driven element is continuous (see impedance matching section). The dipole must be insulated from the boom if it is metal, but the directors and reflectors need not be.

The Yagi designs here are just a starting point for home experimentation. The possibilities for tuning them are endless, but space here is limited. There are some good books available on antenna construction, produced the Radio Society of Great Britain (RSGB) and the American Radio Relay League (ARRL) that are worth reading.

The Log Periodic Array

The log periodic array is similar in construction to the Yagi. This time all of the elements are electrically connected in a zigzag pattern. The result provides an antenna with reasonably flat response over a wide range of frequencies yet is still directional. Commercially available models can be purchased with coverage from 100 to 1,300 MHz within a quite manageable size, or 50 to 1,300 MHz, which is a bit bulkier. The gain of a log periodic is often lower than a given Yagi, but the big advantage of their wide bandwidth is particularly useful in frequency agile spectrometer applications. They are usually used only at VHF or higher frequencies, and construction for the HF bands would generate an enormously unwieldy design that would be difficult to mount and steer.

The design of the log periodic is shown in Fig. 5.8. There are three parameters used to describe the dimensions, α, τ, and σ. The semi angle of the antenna is α. The other two are defined in the formulae below:

$$\tau = \frac{R_{n+1}}{R_n} = \frac{D_{n+1}}{D_n} = \frac{L_{n+1}}{L_n}$$

where all lengths R, D, and L needed to be in the same units.

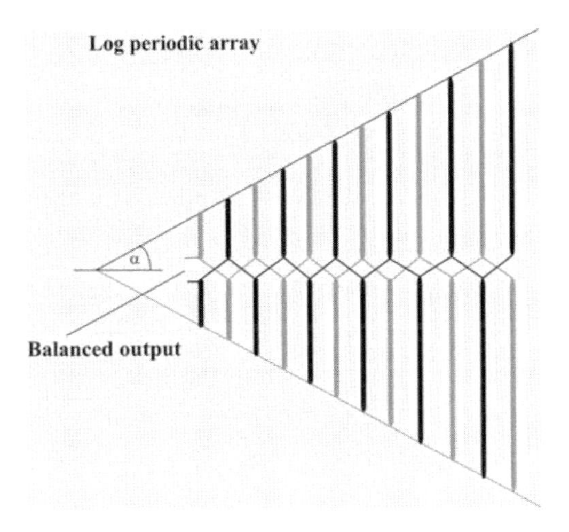

Fig. 5.8. Log periodic array.

$$\sigma = \frac{1-\tau}{4tan(\alpha)} = \frac{D_n}{2L_n}$$

where α is in degrees, and the length units of D and L should be the same.

The value of τ is always less than one and should be kept close to one typically in the range of 0.88–0.95; this leads to the optimal σ value in the range 0.03–0.06. Increasing τ increases the gain and the number of elements required, whereas increasing σ also increases the gain with increasing boom length. Choice of values may be a compromise to yield a manageable size of antenna. The starting point for construction is to decide what the lowest frequency of operation is to be, and for calculation purposes reduce this a further 7%. This will define the longest element width; as with the Yagi, this is calculated from the Yagi equation we saw previously:

$$L_{longest} = \frac{143}{f}$$

where L is in meters and frequency f in megahertz.

The maximum working frequency is usually chosen to be 1.3 times the desired upper frequency.

Let's look at an example. Say we need a log periodic array to cover the band IV/V UHF TV from 470 to 860 MHz. First calculate the longest element by subtracting 7% (33 MHz) from 470 MHz = 437 MHz:

$$L_n = \frac{143}{437} = 0.327 meters$$

Assuming a value of τ of 0.95 then:

$$L_{n+1} = \tau L_n = 0.95x0.327 = 0.311$$

So that:

$$L_{n+2} = 0.311x0.95 = 0.295meters$$

Since the maximum desired frequency is 860 MHz, the upper working frequency is 1.3 × 860 = 1,118 MHz, and the length of the element is therefore 143/1,118 = 0.128 m, Once the chain of calculations above yield an element length of around 0.128 m, then stop.

Now for the element separations:

Choosing a σ value of 0.06 then the spacing between the longest element and the next is:

$$D_n = 2L_n\sigma = 2x0.327x0.06 = 0.039m \ or \ 39mm$$

The next:

$$D_{n+1} = 2L_{n+1}\sigma = 2x0.311x0.06 = 0.037m \ or \ 37mm$$

And continue to the end of the elements.

The length of the elements and their spacing therefore automatically define the angle of the array. It's a bit of a plod through the numbers, but it's worth it if you want to experiment with constructing your own device.

Construction involves a split boom with an insulator between the two layers. The insulator could be air with plastic spacers, or it could be a plastic or wood layer between flat metal strips or metal box sections. Each half of the elements are mounted one on top and one below, and the next pair alternate the opposite way. Larger arrays need to be built rigid, so a box section and round tube or rod is needed. Small arrays for high UHF or microwave use could be built by soldering wires onto a double-sided copper board strip. These arrays can be used at the focus of a dish reflector at microwave frequencies where the elements are very small by virtue of their short wavelength. In this case the elements could be etched onto a double-sided copper board like a circuit.

Circular Polarized Antennae

So far the designs we have looked at have been linearly polarized. Circular polarized antennae are typically used for satellite communication, because the orientation of a satellite is always changing and is often rotating. Satellite manufacturers therefore chose to install circular polarized antennae.

Polarization is seen in natural radio emissions, as was mentioned in the astrophysics sections, although more often there is a random mix of polarizations present. The classical circular polarized antenna is the helical, which looks like a large air-spaced inductor coil. These types of antennae are tricky to make well at home because they involve bending a self supporting conductor accurately, so we won't discuss the details here. The reader is referred to other sources, such as the *ARRL Antenna Handbook*. However there is an easier alternative – the crossed Yagi, the design of which follow the same rules discussed earlier.

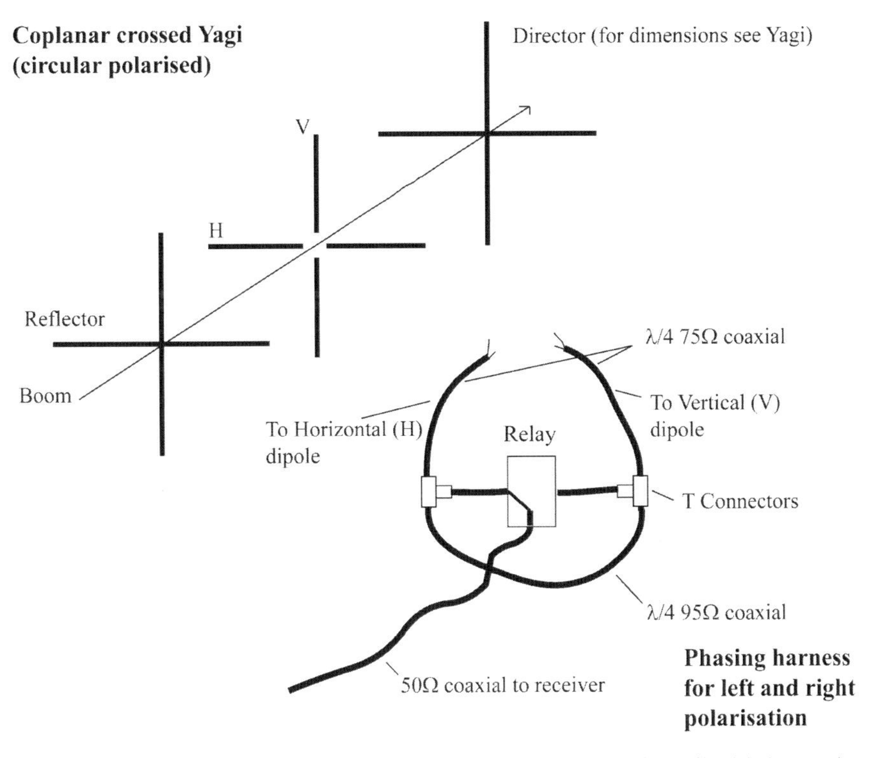

Fig. 5.9. Coplanar crossed Yagi for reversible phasing harness. This style is sensitive to circular polarization and is reversible with the changeover relay.

By mounting a pair of Yagi antennae side by side and 90° to each other they will effectively operate in a circularly polarized manor so long as they are phased correctly together. There are two ways to phase them:

- Coplanar construction, with phasing harness
- Relative offset mounting

The coplanar method means that all Yagi elements for both the horizontal and vertical parts are mounted onto the boom at the same positions, forming a series of crosses. The dipoles are fed by a phasing harness where one antenna has an extra ¼ wavelength of feeder, placing it 90° out of phase with its partner. The diagram shows a method of constructing the phasing harness and of providing a means of reversing the polarization sense from right handed to left handed.

The feed point impedance is affected and requires careful matching due to coupling between the two dipoles. The phasing harness in Fig. 5.9 illustrates how to match the impedance by using coaxial cables of different characteristics.

The Parabolic Reflector ("The Dish")

Now, when you say radio telescope to most people, they immediately think of a large dish antenna. The prevalence of direct satellite reception of television has provided a cheap source of components for the amateur radio astronomer. In practice the dish will be limited to microwave use for the home constructor. Dishes at UHF or VHF frequencies would require large diameters (at least ten wavelengths wide), which are simply not available as off-the-shelf components and would simply not fit in the garden!

Dishes are tricky but not impossible to make at home, but for practical sizes up to about 2 or 3 m they are easily available and are best purchased readymade. The most cost effective way of obtaining a dish for radio astronomy experiments is by searching for used surplus units, for example via online auction sites. It is important, however, to know what the original dish was used for, in order to know what frequency it was designed for. Although a dish will reflect any frequency of radio energy, its physical construction may restrict the upper frequency due to flexure and distortion under its own weight or in wind conditions, and if it is a mesh construction, by the hole size. Many of the large surplus television dishes of mesh construction in the order of 2–3 m in diameter may have been constructed for C band television channels around 4 GHz. Their stiffness may not be good enough to provide a surface accuracy of better than 1/8 wave at Ku band around 11 GHz, although structure could be added to improve this. Mesh hole size needs to be <1/10 wavelength in order that all the impinging radiation is reflected.

The construction of microwave receivers requires considerable skill and is not a task recommended for the beginner, so start out by using off-the-shelf components (see the project chapter devoted to this). With time and experimentation and when you learn more about radio it is certainly an interesting area to explore and offers lots of potential.

The forward gain of a dish is typically in the 20–34 dBi range, and their beamwidth to the half power points can be estimated from:

$$Beamwidth = \frac{57\lambda}{d}$$

where λ is the wavelength in meters and d the dish diameter in meters.

The actual beamwidth will be a factor of how well the antenna is illuminated. Illumination, again referring to a transmitter situation, is how well the antenna and feed horn can utilize the whole aperture of the dish, with minimum overspill.

Now the dish is not an antenna, of course; it is merely a reflector that gathers and focuses radio energy onto the antenna mounted at the focus of the dish. Television systems utilize a device known as an lnb (low noise block). The lnb is more complex than you may realize and consists of at least four elements:

- A feed horn and waveguide
- The antenna, which is a simple straight "probe" (actually two mounted orthogonally)
- An RF amplifier
- A frequency down converter

The feed horn is a collector, which gathers the reflected radiation from the surface. Its diameter is matched to the focal ratio of the dish. In general a short focus dish will have a small diameter feed horn, and the horn diameter will grow with increasing focal length.

Modern satellite TV systems work at frequencies in the range of 10.7–12.75 GHz. If only an antenna was mounted at the focus of the dish, it would need an exotic expensive cable to feed the signal to the receiver, and signal attenuation would be a serious problem. Therefore the signal is first boosted by as much as 40 dB and then reduced to a frequency in the range of 950–2,150 MHz, which is much more manageable and can be transmitted down a fairly inexpensive but low loss 75 Ω coaxial feeder to the receiver. Feeders should always be kept as short as possible. A practical length of up to 10 m will have an attenuation of less than 1 dB. Ku band lnb's can be used for radio astronomy quite effectively, although the average satellite receiver will be useless for this task. The lnb normally gets its power from the receiver, so it is necessary to build a simple device to inject power into the feeder but not the receiver! The project chapters in this book cover this topic.

For the experimenter who wants to try adapting the system to other wavelengths, such as the 21 cm hydrogen emission line, the following guidelines will help as a starting point for the construction of a simple "tin can" feed horn. Referring to the diagram, the critical dimensions are:

- Focal ratio of the dish
- Diameter of the feed horn
- Distance of the probe from the back wall of the "can"
- Length of the probe

The length of the can is not critical, but the probe should not be close to the open end. As a rule of thumb, the can length should be twice or three times as long as the distance from the probe to the back wall. The can is effectively a tube that is closed off at the far end. It acts as a waveguide. At most microwave frequencies, conventional copper cable is terribly inefficient, and "raw" RF signals are directed along metal tubes of very specific internal dimensions. Waveguide technology is affectionately known as plumbing in amateur radio circles (Fig. 5.10).

The antenna is a simple stub, known as the probe. It is essentially a straight wire approximately a ¼ wavelength long. The probe is soldered directly onto the center pin of an N type coaxial socket, which itself is mounted on the side of the horn.

Here we encounter the first problem. Many small dishes are offset types. These are tricky to convert to hydrogen line use, because the feed horn shape is much more complex. It is therefore strongly recommended that a concentric dish be used. Most dishes available in 1.8 m or greater sizes are fortunately concentric. The horn shape can be square, or even oblong, but for this exercise it is recommended that you use a round tubular can. It may be possible to find a metal food can of suitable dimensions. Items such as coffee tins or large cans obtained from commercial catering supply companies have often been used by amateur radio enthusiasts.

Inside the horn waveguide, you can imagine the propagation of radio waves along the tube as a series of zigzag reflections from the side walls, with a reverse reflection from the closed end. The reflected wave comes back up the guide, interfering with the incoming waves. The antenna or "probe," as it sometimes known, is placed at a location where the incoming and reflected waves reinforce each other. A full explanation of microwave waveguides is beyond the scope of this text, but the graphs and formula can be used to gauge the size of the horn and the placement of the probe.

Fig. 5.10. Can-type feed horn for 21 cm hydrogen line.

There is a complication – though isn't there always!? Something strange happens to the wave as it enters a waveguide; the effect is to increase its wavelength, known as the guide wavelength λ_g. All waveguides have a sharp cutoff wavelength below which radio energy will simply not enter the guide. A graph derived from the λ_g equation shows the probe location from the closed end and is of exponential form. The probe position rapidly increases on the left hand side of the graph as the critical cutoff wavelength is approached.

The formula for calculating λ_g is:

$$\lambda_g = \frac{\lambda_0}{\sqrt{1-\left(\frac{\lambda_0}{\lambda_c}\right)^2}}$$

and

$$\lambda_c = 3.42r$$

The minimum horn diameter for a cutoff wavelength of 21 cm is 12.3 cm. Our horn feed must be larger than that. A size of around 15–16 cm should work with most dishes. Probe length will be around 4.6–5 cm long. The probe is usually made adjustable in length. If the probe is a brass rod with a diameter large enough to drill the center and tap a fine thread, then the brass threaded rod is soldered to the center pin of the coaxial socket, and the probe screwed onto it providing adjustment. The exact length can be determined by testing the final system and adjusting for maximum signal. The focus of the dish should lie slightly inside the mouth of the "can style" feed horn. Once again this can be done by testing the final system. The mounting of the feed horn should be constructed in such a way as to provide a small amount of focusing capability (Fig. 5.11).

Simple can-style feed horns obtain most of their signal power from the central part of the dish reflector, where the outer regions are much less efficient. However, they are simple to construct by soldering or welding an end plate onto a metal tube of suitable diameter. There is nothing wrong in starting experiments with such simple devices.

An improved feed horn has a choke ring mounted on the open end. The choke ring should be made to slide up and down the cylindrical horn for fine tuning. This improves the "illumination" of the dish and improves its efficiency around the periphery.

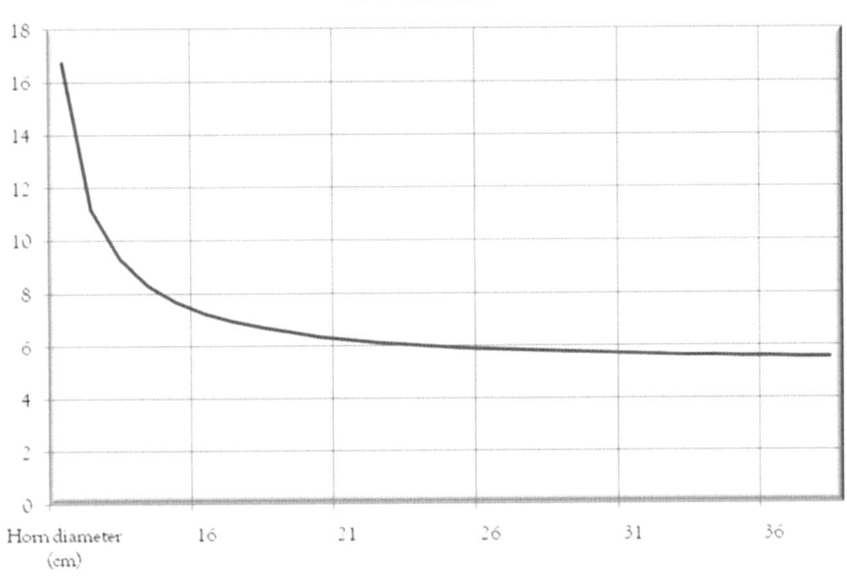

Fig. 5.11. Feed horn at a dish focus. The vertical axis of the graph represents the position of the probe from the closed end of the horn for a wavelength of 21.1 cm (the cold hydrogen emission line), and the horizontal axis is the diameter of the horn. The cutoff wavelength is around 13 cm at the left hand side of the graph. Clearly for a diameter of less than 16 cm the position moves quickly as the cutoff is approached.

Table 5.1. Dimensions of a typical 21cm horn feed

Frequency	1.42 GHz
Wavelength	21.1 cm
Horn (waveguide) diameter	15.2 cm
Lower cut off frequency	1.14 GHz
Upper cut off frequency	1.49 GHz
Waveguide (Guide) wavelength	35.2 cm
Probe placement from closed end	8.8 cm
Feed horn length	26.4 cm
Choke ring depth	10.6 cm
Choke ring diameter	36.4 cm
Dish focal ratio	0.4
Dish focal point depth inside horn	3.7 cm
Distance from front of horn to back wall of choke ring for minimum noise	10.4 cm
Distance from front of horn to back wall of choke ring for maximum gain	11.7 cm

To summarize the dimensions and design of a 1,420 MHz feed horn assembly for HI emission line work refer to the Table 5.1 and diagram. The dimension should be good for a dish with a focal ratio of between 0.25 and 0.5 (Table 5.1).

Antenna Impedance Matching

Correct impedance matching is essential to efficient operation, especially considering the weakness of the signals encountered in radio astronomy. Mismatched impedances at any point in the antenna/receiver chain will create internal reflections, reducing the amount of power passed on to the next stage.

Before we look at methods of impedance matching we need to take time out and look at feeder cables, connectors, and their properties.

Transmission Lines (Feeder Cables)

It would be wrong to call a feeder cable a "wire"; it is not just a wire. You must think of it as a component in the circuit, or even as a circuit in its own right. It will have inductance, capacitance, and resistance properties. If the feeder was only carrying DC signals, then only the resistance would matter, and that is very low for short runs. Dealing with RF signals is quite different. Capacitors and inductors can act to impede the flow of radio frequency signals in the same way resistance impedes the flow of DC signals (as well as AC signals).

For the purposes of RF antenna connections there are three main types of feeder constructions available:

- Twin lead, consisting of a pair of insulated parallel wires separated by a plastic sheet spanning the gap. The sheet may have regular holes in some types.
- Parallel open lead, similar to the twin lead, but uses plastic spacers to keep the wires apart, with much more free space between them.
- Coaxial lead, consisting of a central insulated wire, surrounded by a copper braided sheath which itself is insulated on the outside. Some types have an additional copper foil layer under the braiding to improve high frequency performance.

Twin lead is most often encountered with a characteristic impedance of 300 Ω used for television applications, fitted to the feed point of a folded dipole. It can also be found with an impedance of 450 Ω, where there are usually a regular series of holes cut in the insulation between the cores. The main consideration when using twin feeders is the routing of the feed must not pass close to metal structures, especially earthed metal structures, or anything likely to conduct electricity to earth.

Parallel Open Lead

This type is similar to twin lead. I consists of a parallel pair of wires, with regular insulating bridges between them to maintain a constant spacing. The impedance values available span the range from 300 to 1,000 Ω. The impedance being defined by the diameters of the conducting cables, and the spacing between them, greater spacing provides higher impedance values.

Coaxial Cables

Coaxial refers the concentric design, where one conductor is surrounded by another. They are available with impedance values in the range of 36–120 Ω, but 50 and 75 Ω values are most common. It is conventional, but not mandatory that 50 Ω (such as RG 58/U) is used in radio circuits, and 75 Ω (such as RG 59/U) in video and television circuits.

The many types differ in diameter, and in the type of insulators used. Insulators between the conductors are often polyethylene, polyfoam or Teflon and in some types there are air spaced channels running the length of the cable. It is important that connectors and connector joints that are exposed to the weather are carefully sealed against water ingress. Water in the cable center will change the dielectric properties seriously. A special tape known as self amalgamating tape is good for this. The tape is first stretched and then wrapped around the joints. As the stretched polymer relaxes it binds to the layers of tape above and below making an effective seal without the aid of glue which tends to dry out and come loose eventually.

One important property of transmission lines is their velocity factor. Velocity factor is denoted by V or VF, and is the ratio of the velocity of a signal in the cable, over the velocity in free space. Radio signals travel at the speed of light in open space, but the same signal captured by an antenna and converted to an electrical signal which travels more slowly in the feeder. Therefore the velocity factor is always less than one. The table gives the typical values for some cable types.

Cable type	Velocity factor
Parallel open line	0.95–0.99
300 Ω twin lead	0.82
300 Ω twin lead with holes	0.87
450 Ω twin lead	0.87
Coaxial cables	0.66–0.80
Coaxial with polyethylene insulator	0.66
Coaxial with polyfoam insulator	0.80
Coaxial with Teflon insulator	0.72

Fig. 5.12. Common RF connectors. From *top left*: BNC plug, BNC socket, PL259 plug and matching SO239 socket, UHF "TV" plug, N plug, N socket. From *bottom left*: F plug, F socket, SMB plug, SMB socket, SMA plug and SMA socket. Note that most types come in variations such as straight or angled and are available for different sizes of cable. There are also 50 and 75°Ω impedances in some types. SMA plugs come in normal and inverted pin types.

The velocity factor is important when you come to calculate precise lengths of cable in units of wavelength for making matching sections, cable baluns, or phasing harnesses. For example, the length of a half wavelength phasing section often referred to as the electrical length, made from polyethylene coaxial cable, is 0.66 times shorter than it would appear based on the actual wavelength. For a 2-m radio signal wavelength, a half wavelength is 1 m, but if this signal is carried in the above coaxial cable, the electrical length representing a half wavelength is 0.66 m or 66 cm!

Modern practice prefers the use of coaxial cables as feeders. These need to be connected to the receiving equipment using a convenient plug and socket arrangement. The most common connectors encountered are BNC, RCA (phono), N-type PL-259 (which mates to a So-239 socket), SMA, and SMB. BNC's are best avoided for RF use but are fine at low and very low frequencies. Although the RCA was originally introduced as a UHF connector, it is most often used as an audio connector now. The N-type is preferred for UHF applications but will work well at low frequencies, too. The SMA and SMB are very small and are used for UHF to microwave frequencies. If you like to salvage components from junk equipment, be aware that connectors also have characteristic impedances, again typically 50 or 75 Ω. Some, like the N-types, are not always compatible if you mix socket and plug impedances (Fig. 5.12).

Impedance Matching

Amateur radio enthusiasts often employ an ATU (Antenna Tuning Unit) to match the impedance of antennae to the receiver or transmitter. These come in a variety of shapes and sizes, and some are dedicated to a particular transceiver. All ATU's will only work efficiently at specified wavebands. Commercially built units will almost certainly be optimized for specific communications wavebands. They may not be suited to the frequencies required for radio astronomy. It is always preferable to construct the antenna to be a good match to the receiver without the need for intermediate variable tuners. However, even for a fixed frequency radiometer, some provision for matching is built into the system.

The Yagi antenna, sometimes known as a " beam," can be constructed slightly differently using a technique known as a gamma match. This provides adjustment for the feed point impedance. We saw that the feed point impedance of a dipole mounted in a beam is less than a dipole on its own and will be in a range typically from 20 to 36 Ω. If the impedance was as high as 36 Ω that would be considered a reasonable match to a 50 Ω coaxial cable, but could be improved using a gamma match system. Here the dipole is no longer split into two parts, or folded. The drive element is now a single rod of a length equivalent to that of the unfolded dipole. Note that this should be a self-supporting solid rod or strong tube. Alongside this is mounted a shorter rod of the same diameter, with one end mounted at the center point of the driven element a small distance away. An adjustable shorting link connects the short gamma matching rod to the driven element, and therefore "taps off" the driven element off center. The tap-off point is adjustable. The shorting link is made to slide along both the driven element and the gamma matching rod. Figure 5.13 shows the connection details and layouts. The formulae for calculating the dimensions are:

$$G = \frac{D}{10}$$

$$S = \frac{D}{70}$$

where D, G, and S are measured in the same units.

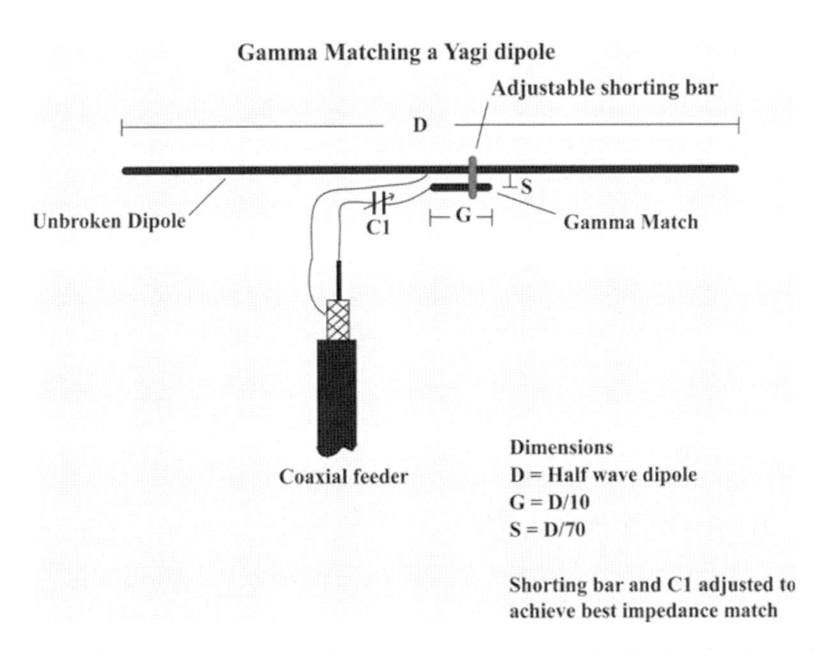

Fig. 5.13. Gamma match. Note that the driven element is no longer split in the center. It makes this an ideal addition for Yagi driven element matching.

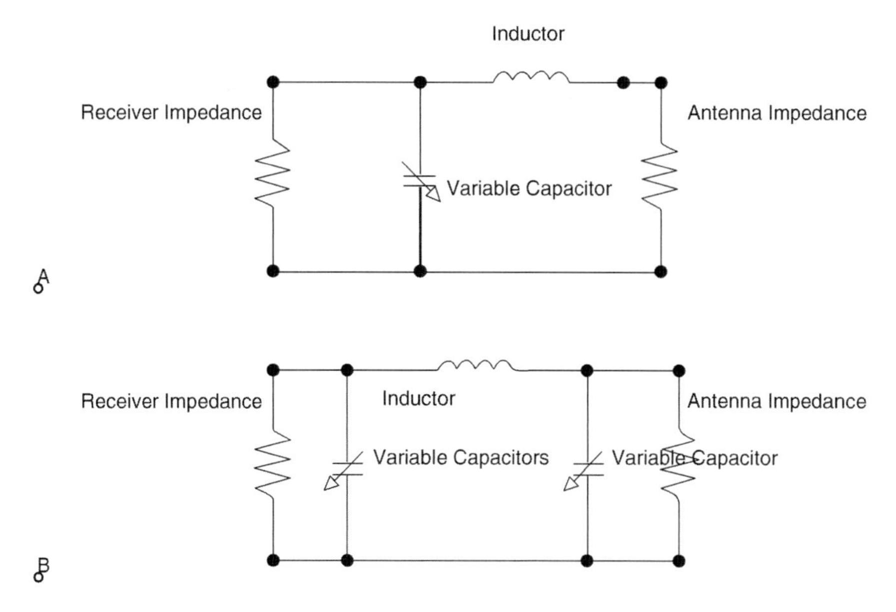

Fig. 5.14. Antenna matching networks.

The easiest way to set up an antenna like this is to use an antenna analyzer, such as the MFJ-269 manufactured by MFJ, although this is only valid for the frequency ranges 1.8–170 MHz, and 415–470 MHz. Alternatively use an impedance bridge, as we shall see in Chap. 8.

To adjust the impedance of an antenna, a capacitor and inductor may be used. The capacitor is usually made to be variable. The old-style air-spaced tuning capacitors are useful for this task but are becoming increasingly difficult to buy new now, although many can still be found at radio rallies salvaged from old equipment. An alternative would be to use a bank of fixed value capacitors and a series of dip switches to add or remove value from the stack. The inductor components in tuners are usually air-spaced again and can be constructed with "tap off" wires on some of the coils; so by using a rotary wafer switch longer or shorter sections of coil can be selected, thereby varying the inductance value.

Figure 5.14 illustrates two ways of connecting these components to an antenna. They are used where the receiver impedance is greater than the antenna, such as the case of the Yagi, where the feed point impedance is often less than 37 Ω.

Typical values of the capacitor are 140–250 pF and the inductor 18–28 μH.

Using Coaxial Cable as a Matching Unit

By design our receiver will have a standard input impedance of say 50 Ω, so we would feed it with 50 Ω coaxial cable. However, the antenna feed point, as we know, may well be different. It is often possible to use a quarter wavelength of coaxial cable of a different impedance to match the two, so long as the required value is available from the stock lists.

In order to calculate the details we not only need to work out what impedance value is needed but how long the section needs to be. Don't forget the electrical length of one quarter wave depends on the velocity factor of the cable used.

The following formula is used to calculate the impedance of the matching section:

$$Z_M = \sqrt{Z_L Z_F}$$

where Z_M is the impedance of the matching section, Z_L is the impedance of the antenna feed point, and Z_F is the impedance of the main feed line.

For example, if the antenna feed point impedance was 120 Ω and the main feeder is 50 Ω, then Z_M is close to 77 Ω. Therefore one quarter wavelength of 75 Ω feeder would be an excellent match. If this was operated at a wavelength of 4 m the velocity factor of RG59/U would be about 0.82, so one quarter wavelength is $4 \times 0.25 \times 0.82$ m, which is 82 cm.

If the feed point impedance was 28 Ω, then the matching section would need to be close to 37 Ω. Belden 8700 has an impedance of 32 Ω, which would be a fairly good match, but it is just over 1 mm diameter, not very practical, so another method of matching would be better.

Matching Stubs

The quarter wavelength stub is a useful way of matching a dipole antenna. Here a quarter wavelength of parallel transmission line is attached to the center feed point of the dipole. The conductors should not be insulated so that a tapping point can be found along the length of the stub that matches the impedance to the coaxial feeder. The parallel conductors could be made a structural part of the antenna and conductive clamps attached to the stub. The coaxial is connected to these via a 4:1 balun. Impedance matching is performed by moving the clamps up and down the stub. The end of the stub has a shorting bar across it that is also adjustable. Refer to the diagram for an illustration. The usual way of calculating the dimension L1 and L2 involves measuring the VSWR (Voltage Standing Wave Ratio), but this assumes the antenna is connected to a transmitter. For the purposes of radio astronomy the adjustment of the feed point and shorting could be done by trial and error to maximize a received signal close to the operating frequency, but this is likely to prove impossible. Once again refer to Chap. 8 for the discussion on impedance bridges. These really help to take the mystery out of the art of matching (Fig. 5.15).

Matching Transmission Lines to Multiple Antennae

One device used in radio astronomy is the interferometer. By using two or more antennae separated by at least a few wavelengths, higher resolutions can be achieved. Or simply by adding a pair of antennae their collecting area is effectively doubled.

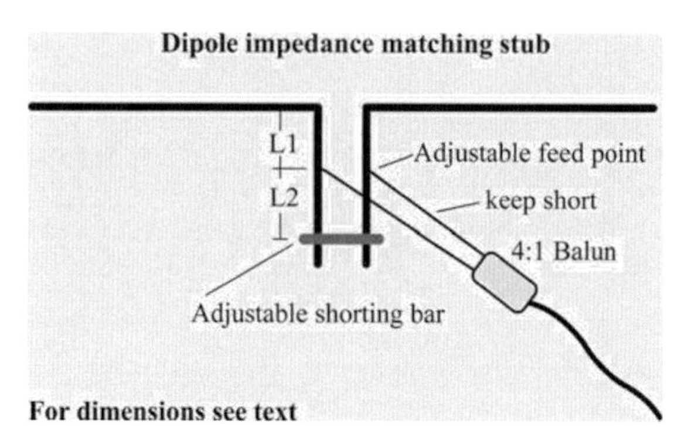

Fig. 5.15. Dipole with matching stub and variable feed point.

Since only one receiver is used, the two antennae need to be correctly matched for impedance. If a pair of 300 Ω twin line feeders was simply joined to one 75 Ω line, there would be a big mismatch, and a lot of signal power would be reflected back to the antennae at the joint. The pair of lines from the aerials is effectively in parallel, so their combined impedance will be 150 Ω. We need to transform the 150–75 Ω. This could be done with a quarter wavelength section of transmission line having an intermediate impedance of the geometric mean of the two parts. (The geometric mean of two numbers is $\sqrt{Z_1 Z_2}$). So if we could find a feeder with 106 Ω characteristic impedance then we are in business - hmmm, maybe not.

Well, it turns out that if we use a parallel conductor air-spaced feeder, we can manufacture a quarter wavelength section with any impedance value in the range of 210–700 Ω. That does not appear to help.

Let's now rethink the feeder scheme for a minute. If we have chosen to use a pair of antennae with a feed point impedance of 300 Ω, then the 300 Ω twin line feeder is easily obtained to interconnect the antenna pair. At the central point on the interconnection we can tap off a feed to the receiver, so that both antennae see exactly the same length of line to the back end. The new feed point impedance is 150 Ω. If we now match that to another length of 300 Ω twin line, then we need a matching section of 212 Ω feeder. The table shows how the values of impedance vary with separation for parallel air-spaced conductors.

This means that if we mounted a pair of 6 mm tubes rigidly with a separation of 1.9 cm (interpolating from the table), we can make a quarter wave transmission line. The air-spaced parallel line has a velocity factor of about 0.95, so that the length of the tubes will be 0.95λ. Plastic insulating mounts should be fabricated to hold the tubes.

The one thing to remember is that parallel lines must be kept away from structures, particularly metal ones. The advantage is a parallel feeder has lower loss than most coaxial lines. The feeders should therefore be suspended above ground. Once they reach the building housing the receiver then a 4:1 balun can be installed (see next section), which reduces the impedance to 75 Ω, and a short piece of 75 Ω coaxial line can be used to enter the building and connect to the receiver.

Table of impedance for open-air dielectric parallel transmission line

	Center spacing (cm)				
	1.25	2.5	5.0	10	15
Wire gauge					
20	420 Ω	500 Ω	580 Ω	660 Ω	710 Ω
16	370 Ω	440 Ω	515 Ω	610 Ω	650 Ω
12	300 Ω	380 Ω	460 Ω	530 Ω	590 Ω
8	240 Ω	320 Ω	410 Ω	480 Ω	530 Ω
Tube diameter (mm)					
6	157 Ω	230 Ω	330 Ω	410 Ω	460 Ω
12		157 Ω	250 Ω	330 Ω	380 Ω

Baluns

A balun converts a balanced unearthed antenna such as the dipole to an unbalanced system such as an earthed coaxial receiver input. A simple 4:1 balun required in the stub matching system described above can be made from a half wavelength of coaxial feeder of the same type used to connect the antenna to the receiver. The effect of this also reduces the impedance to a quarter of its original value. If our receiver is made to have a 75 Ω input impedance, then the combination of a dipole with a quarter wavelength stub and a coaxial 4:1 balun will provide a good match to a 75 Ω coaxial feeder (see Fig. 5.2).

The coaxial balun is only good at one frequency, so it would be suited to a fixed frequency radiometer. If a wider bandwidth is required for a tunable receiver or spectrometer, then a different scheme is required. Wideband baluns can be made using ferrite or powdered iron toroids.

The two most common styles of balun are shown schematically in the diagram. Style A is a 1:1 balun and offers no impedance change, while style B is a 4:1, which reduces the feed point impedance to a quarter of its original value. The coils are all wound together onto a ferrite or powdered iron ring. The two or three coils are made up from two or three strands of enameled copper wire laid side by side and wound as if they were one wire onto the toroid. This is known as bifilar or trifilar winding (Fig. 5.16).

The toroidal powdered iron and ferrite cores can be used to build baluns. At the same time they can be used to transform any practical impedance to match the receiver. There is a section in the introduction to RF electronics that gives a method of calculating the number of turns on each winding to match the impedance. Note that in the standard 1:1 and 4:1 versions illustrated here the number of turns on each winding is equal. When dealing with other impedance ratios the number of turns on each winding should be different. A toroidal balun transformer is a wideband transformer but is still somewhat frequency dependent. The balun should be wound for a particular center frequency of interest.

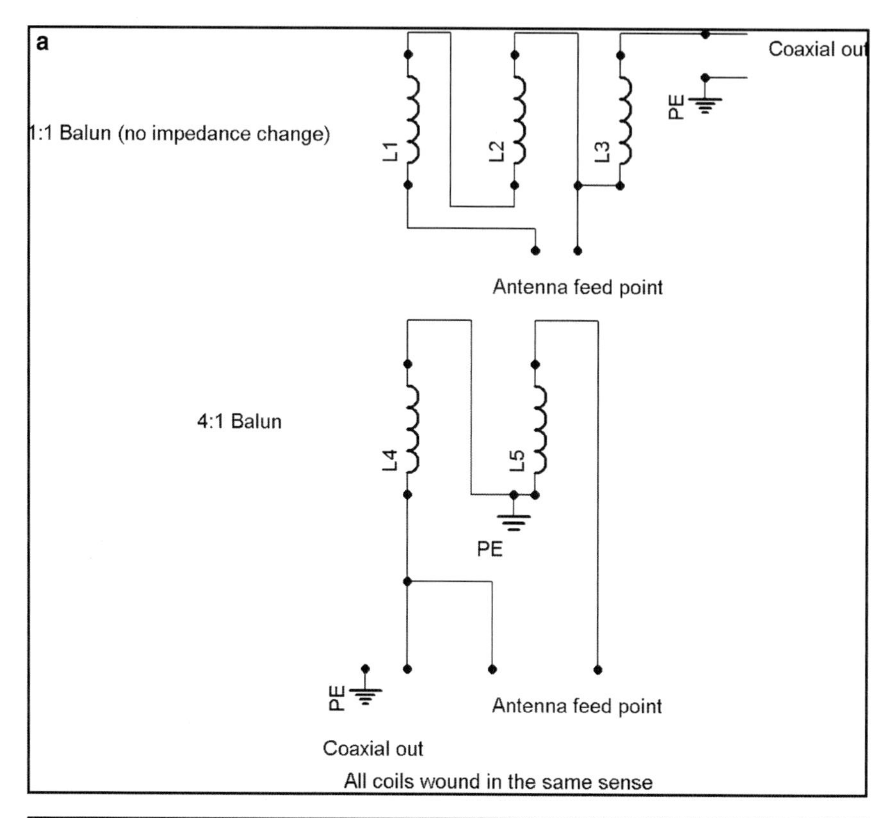

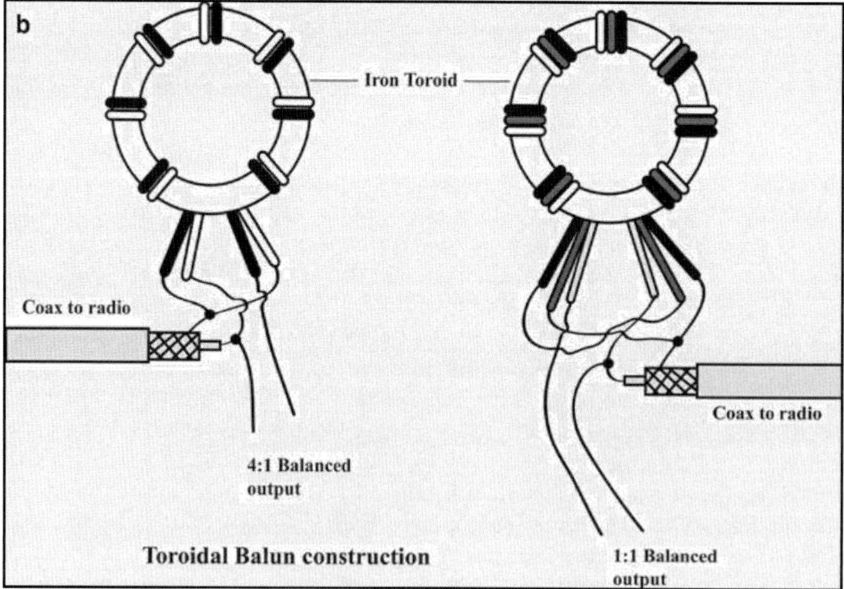

Fig. 5.16. (**a**, **b**) Toroid baluns 1:1 and 4:1, and toroidal winding style.

Chapter 6

Setting Up a Radio Astronomy Station

Selecting a Site for the Receiving Station

For most observers site selection will be an easy choice – your home. Although you may not have any choice of alternate locations, some consideration has go into the practicalities of constructing a receiving station. You need to have space for the antennae you intend to use. Some aerials need lots of room, others can be quite compact, so depending on the space you have available this may determine what kind of radio astronomy is possible.

Very low frequency propagation studies as a way of monitoring solar flares should be possible anywhere. The small loop antenna even works well indoors and does not take up a lot of room. However, setting up an interferometer to work at 80 MHz using an array of four dipoles may well require more space than is available in your garden. In general aerials get smaller with increasing frequency, so working at higher frequencies is better suited to small spaces, although above 1 GHz a dish reflector is usually needed to increase the collecting area and improve the efficiency for detecting weak signals. Although the antenna is very small, the dish needs to be large to increase its collecting area.

The alternative to your garden is to find a suitable large site nearby. This may be easier said than done, and it may involve having to pay rental of some kind. Astronomical societies working as a team could construct a radio observatory in a more remote place. Being away from built up areas also helps to isolate the receivers from some types of radio noise. However, be aware of radio towers nearby used for things like pagers or mobile phones. These will cause problems in some frequency bands. Mobile phone masts in the UK operate at around 872–960, 1,710–1,876, 1,920–2,169 MHz. Although these are not the only strong signals to avoid, the masts tend to be spread all over the place and could be right on your doorstep.

When evaluating a site try to survey it with a scanning receiver. It will tell you what sorts of signals, or noise, is present in the region. Sources of noise can come from strong transmissions at frequencies adjacent to the chosen one. Choice of radio channel for forward scatter meteor astronomy requires that no local

J. Lashley, *The Radio Sky and How to Observe It*, Astronomers' Observing Guides,
DOI 10.1007/978-1-4419-0883-4_6,

station is present, yet a distant transmitter in the range of 500–1,500 km should be available on that channel. Ideally the neighboring channels higher and lower should also be clear locally. When relying on the broadcast FM band this could be challenging at best, or maybe impossible.

Other sources of interference could be computers, including the one you will use to log the data. Metal-cased computers may be better for radio astronomy. The casing will be grounded, reducing the chance of stray noise in the locality. Clearly the covers should always remain in place. If interference is tracked to the PC, then keeping it as far from the radio receiver as possible should help.

Radio equipment needs to be grounded with a purpose made grounding rod. It is preferable to place the grounding rod into the earth outside. Damp ground is better than very dry soil. Clearly this is not possible for apartment dwellers. A substitute could be a water pipe or similar device where these are metal.

Electrical noise can be a problem, especially near industrial sites where large electrical machines are operated. Many residential properties are isolated from such industrial locations, but be aware of them. Cars, too, were traditionally quite noisy, generating spiky noise from their ignition systems, but they are much better now than they used to be. Other forms of radio interference can be found from amateur radio bands, CB, and television. Of course, you need to operate away from these channels, but adjacent channel interference can still be a problem and does not necessarily enter the system through the aerial device. Direct pick up into various parts of the receiver is possible, but a receiver well shielded in a grounded metal box will help stop this. Interference can even enter the radio via the power connection. However, well made modern stabilized power supplies should filter that out. Even house lighting can cause problems. Modern fluorescent low energy lamps do generate some RF noise. Some lights are clearly visible in the spectrum of the VLF receiver. Old style CRT television is also strong at VLF frequencies, particularly 15,625 kHz.

Atmospheric Noise and Other Environmental Considerations

The strongest noise from the atmosphere is caused by electrical storms. Some areas of the world suffer worse than others in this respect. It is wise to shut down receivers during local electrical storms and isolate the antenna feeders, ensuring that they are grounded. Although a direct hit of lightning will be disastrous, induced currents by nearby strikes are more likely and could still be strong enough to destroy sensitive radio equipment.

Electronic equipment is designed to work best at room temperature. Strong swings in temperature can seriously affect system performance. Also large swings in system temperature can significantly change the noise performance of radio systems. Colder temperatures will reduce system noise somewhat but may make comparison of observations from different seasons a problem. The main receivers and computing equipment should ideally be kept as close to a constant temperature as possible. This means working in a garden shed environment a bit more challenging. At least evaluate what different temperatures do to the performance of the system. It's not that you can't work in a garden shed, but just be aware of temperature effects.

Antenna Mounting

A number of considerations must be given to mounting aerials. Wind loading is one. A large solid dish antenna clearly presents a hazard in strong winds. Mesh dishes are only slightly better in this respect. The pole on which a dish is mounted needs to be strong enough to withstand up to 100 mile/h winds. This author once had a dish only 1 m across, mounted on a 2-in. steel pole that admittedly was a thin wall type, but it was provided with the dish mount. In a strong windstorm the pole was bent over 90°! A thick-walled steel or aluminum pipe such as a scaffold pole would have survived that storm.

Even open-wire dipoles may be susceptible to wind damage if the wire it is made from is too thin. Strong rope supports are needed at the tips of the aerial, and dog bone insulators used between the wire and the rope. Choose porcelain insulators over plastic ones (Fig. 6.1).

The center point of dipoles should be strengthened with a insulating material. If the antenna is kept reasonably taut but not over tight, it will reduce its flapping

Fig. 6.1. Porcelain "dog bone" insulators.

tendency in winds, which lead to fatigue in the wire and potential breakage. If the aerial is over tightened in summer, then thermal contraction in winter may be enough to snap the wire.

Certain antennae need to be kept up in the air a minimum distance from the ground to avoid their impedance properties being affected by Earth.

If a mast is required to support antennae, it will need a good footing, usually concrete, in the ground. The stability of the location should be considered. If it is too sandy, or too wet and soft, it is not a good location for mounting a tall mast.

Power Considerations

If you are working from a fixed location with the availability of mains power there should be no problem with power stability. The receivers described in this book all work from low voltage and should be powered from commercially purchased regulated DC power supplies. There is no need to construct your own power supplies from scratch, and this should certainly be avoided if you have no experience in this area. What you need to be aware of is that not all of the little black box power supplies are in fact regulated. An unregulated 12-V power supply is often used in battery chargers, and the output voltage will be more than the face value at 14 or even 15 V. Unregulated supplies will not be stable under variable load conditions either. Always use a power supply that is marked as regulated for radio use. In fact many radio projects use internal regulator chips to guarantee a stable power source. It is wise to do the same in your projects. However it means that if you require a 12-V supply, and you use a 12-V regulator in your design, then you will need to connect a power supply of a slightly higher voltage, say 14–16 V. Most regulator ICs have a wide tolerance for input voltage, but it must be higher than the voltage it is expected to output. Check its data sheet for information.

If you are working at a remote site without power or in a mobile situation then you can't beat a good battery as a power source. For example a car battery will provide many hours of power for low-power receivers, or for extended use between charges a leisure battery as used in caravans is best, but more expensive. However, computers used for data logging can be a big drain on battery power. This is where the dedicated microcontroller data logger is a useful tool. Refer to Chap. 13 in this book for details of such devices. If long-term remote monitoring stations were contemplated, then consider installing a solar battery charger.

Chapter 7

Radio Hardware Theory

The essence of this book is about learning the concepts of radio design, not just copying a recipe for success. This chapter is all theory about how radios work. Once you have an understanding of the principles, you should be able to begin experiments of your own, not necessarily copying a design but modifying a design to work with different components, or to work on a different frequency, for example. Building a known design to start with and getting it working gives you a boost in confidence to carry on and go further. Modifying a design or designing your own gives you a real sense of achievement.

Before looking at how a radio works, the question you may be asking yourself is, "*Why do I have to construct anything? Aren't there lots of radios I can buy ready built?*" It is true there are lots of radios out there, and many could be used for some radio astronomy experiments. But most have disadvantages for radio astronomy, such as narrow bandwidths, automatic gain controls, etc. These features will be covered in the discussions to follow.

A radio telescope is a simple device; often its only task is to measure signal strength, or more exactly to measure received noise power. Usually this involves reducing a high frequency radio signal to a lower, more manageable, frequency and taking the power measurement there. There is no need for complex demodulators and fancy features found in modern receivers. A basic radio telescope will operate at a single frequency (actually a band of frequencies) and may have very few user controls. Even the simplest devices can still be quite effective. Amplification and noise rejection are its main roles. The "signals" received from natural objects such as Jupiter and the Sun are not talking to us, and there is no encoded data. Note I placed the term signals in quote marks to emphasize a point. The definition of signal implies a message is being communicated. This is clearly untrue of natural space radio sources, but because we can study the physics of natural processes by analyzing radio sources, in a twisted way there is information to be had so I will use the term signal in this context. The alternative is to refer to these "signals" as radio noise, but this would be confusing. As astronomers we want to measure the natural radio noise from space but discard the noise generated within our receivers. Not an easy task at times!

The Superheterodyne

To begin looking at how radios work we will concentrate on the superheterodyne. There are simpler designs, but the superhet, as it is often known, is by far the best

J. Lashley, *The Radio Sky and How to Observe It*, Astronomers' Observing Guides,
DOI 10.1007/978-1-4419-0883-4_7,

for our purposes. The word "heterodyne" refers to the way two signals are mixed together to form a new frequency, or beat frequency, which is usually but not always lower in frequency than the radio channel.

It is helpful to break down the design of radios, or any electronic circuits, into blocks or modules. These blocks are self contained and could be built independently of the rest of the system. In fact it is recommended to do it this way. By building a radio as separate modules, possibly even in separate screened boxes, it makes it much easier to modify the system later to, say, change the frequency range it covers, or to improve its rejection of unexpected out of band interference, etc. The block diagram is shown in Fig. 7.1.

The items to the left of and including the mixer are known as the "front end." The filter and the mixer are usually combined together into the same board or housing. The filter is needed to remove unwanted mixer products.

The RF amplifier's job is to boost the signal received by the antenna and is often mounted as close to the antenna as possible. There may or may not be an additional stage of amplification before the mixer within the main receiver. A broad band RF amplifier is known as a preamplifier, but in practice filters may be combined with the RF amp to reject unwanted out of band signals interfering with the system. It is then better described as a preselector.

In more sophisticated radios the RF stage may include tunable filters in the preselector that track the required frequency as the system is tuned. Making tunable front ends is beyond the scope of these projects, but it is relatively straightforward to build fixed filters to preselect a required band. Filter design will be covered later.

The mixer stage involves combining the radio signal with an artificially generated sine wave from the local oscillator (LO). Mixers are non linear devices. They generate output frequencies that are the sum and difference of the RF and LO signals. Mixers that are unbalanced also contain copies of the RF and LO as well as their sum and difference. However, double balanced mixers are recommended because they significantly attenuate the RF and LO signals at the outputs.

As an example, a double balanced mixer fed with a radio frequency of 22 MHz, and mixed with a LO of 21.5 MHz will output signals at 500 kHz and 43.5 MHz. If it

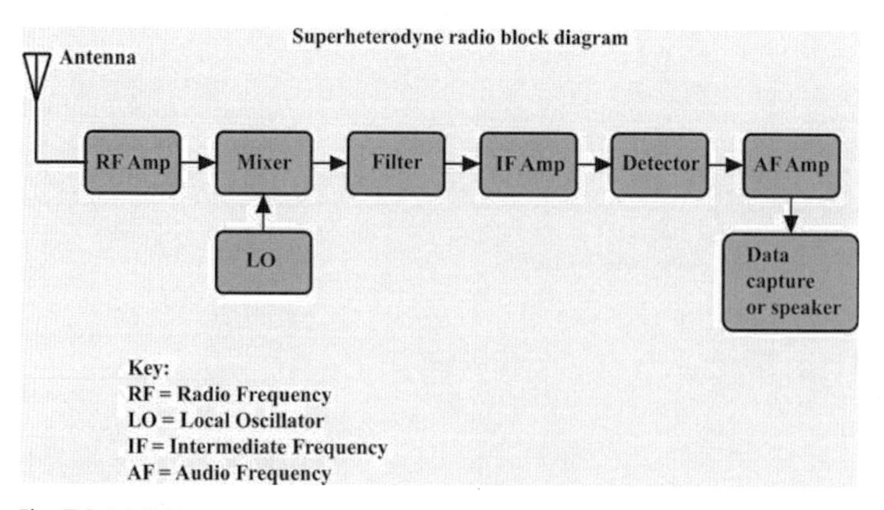

Fig. 7.1. Radio block diagram.

is the lower output that is required the filter following the mixer should be a low pass variety, which blocks the 43.5 MHz product and the original 22 MHz. The actual output of the mixer is more complicated because the local oscillator will produce overtones of its operating frequency at integer multiples of its design frequency. These will also be mixed with the RF signal. So a band pass filter is more often used after the mixer stage to isolate only the output that is required, and attenuate the rest. Even adding a band pass or low-pass filter to the output of the oscillator to select its design frequency may be a good idea to keep the mixer output clean.

At the filter stage it is important to remove unwanted products from the mixer. Practical filter design will be covered later. Refer to Fig. 7.2 to understand the effects of different filters. Filters come in three main types: band pass, low pass, and high pass. Band pass filters allow a selected range of frequencies through to the next stage. Low pass allow low frequencies through while blocking higher frequencies. The high pass is the opposite of this. Filters are designed by deciding where the cutoff must be, and which frequencies need to be blocked. The cutoff points are usually defined as the -3 dB points. This is where the power falls to half of its maximum value. The degree of attenuation at a given frequency is also designed into the filter. In the diagram examples this was set at -40 dB. The difference between this and the −3 dB is known as the stop band. The frequency at which the −3 dB points and stop band points occur define the slope of the filter. If the side slopes are steep the filter is more complex and challenging to build. If the slopes are shallow the filter is simpler and much easier to fabricate.

The IF, or intermediate frequency, amplifier usually does most of the amplification. The mixer is used to reduce the radio channel to the fixed IF. This method makes it easier to design stable high-gain amplifiers that only have to deal with a narrow range of frequencies that don't have to be tuned. It is also much easier to build filters of high quality if the frequency is fixed. Another role of the IF amplifier is to define the band width of the receiver, and to actively reject all those frequencies outside the desired range. In some cases it is desirable to use several different filters in the IF stage, providing a range of receiver band widths. A multi-position switch is used to select the desired one.

The output of the IF unit is a copy of the radio channel, which is translated into frequency. In communications receivers this signal contains a carrier frequency and/or modulated sidebands containing audio information. The detector separates the two, removing the unwanted carrier, leaving an audio signal that can be amplified and heard. In radio astronomy, of course, there is no modulated signal, although the detector can still provide an audio output that will vary with the change in amplitude of the received signal in the same way. It is this variation in amplitude we need to measure, record, and analyze. When dealing with natural radio sources we are in effect measuring the changes in the noise background, which for many objects sounds like the classic hiss. As we shall see later, an alternative to the traditional detector is a logarithmic amplifier. The output of a log amp is a DC voltage whose magnitude is directly proportional to the power of the signal, measured on a decibel scale.

The audio amplifier would follow a traditional detector. In a basic radio telescope the detector would usually be a simple diode, making it effectively an AM or amplitude-modulated receiver. The role of the audio amplifier is to boost the power to the point at which it can drive a speaker. In practice the speaker is not necessary and would be replaced with an integrator or square law detector, which provides a DC output voltage proportional to the received signal amplitude.

This completes the description of a basic receiver suitable for radio astronomy projects. It should be noted at this point that most commercially built receivers include an automatic gain control, or AGC. This provides a voltage derived from

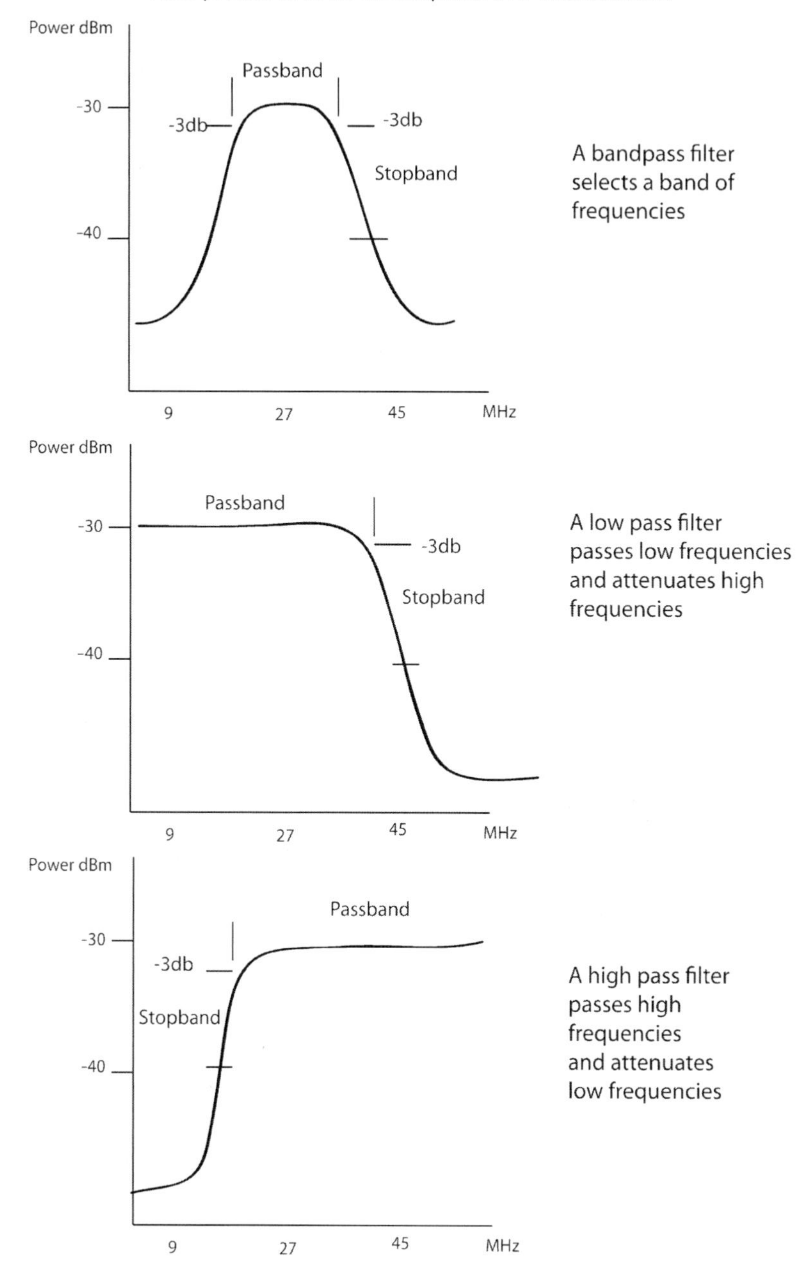

Fig. 7.2. Filter profiles.

the IF stage that controls the gain, or amplification of the RF stage. This is an undesirable feature in a radio telescope. AGC is provided to help smooth out the effects of fading of long-distance communications signals, and to provide constant output level when working between strong and weak channels. Clearly in radio astronomy we are most interested in the difference between strong and weak sources, and in any changes that occur in their strength. Therefore the AGC system should be omitted from the design, and only manual gain control used. Commercial receivers chosen for use in radio astronomy should have the ability to disable the AGC in favor of manual adjustment. A feature of variable bandwidth would be good, too, from narrow – a few kHz – to at least a few tens of kHz.

In the discussion so far we have not mentioned anything about the difference between an AM and FM radio. You will be familiar with these terms from the range of broadcast band receivers on the market. The superheterodyne principle works for all modulation types. Amplitude modulation is where the amplitude of the radio wave is varied and carries the audio information. An FM signal has constant amplitude, but the variation of radio frequency carries the desired audio. It is only the detector stage which differs. FM receivers will not be discussed here, as they are unsuitable for radio astronomy use (except for basic meteor scatter work, which we saw earlier). Other modulation types include single side band (SSB), used by amateur radio enthusiasts. See the Chap. 3 for one use of SSB receivers.

Measurement Scales

When designing or buying a radio for astronomy use there are a number of things to consider in their specifications. Next we will look at the requirements in detail. Before that it is important you understand some of the units of measure used in radio specifications.

Radios need to be very sensitive, in order to handle the tiny input voltages obtained from an antenna. These voltages are measured in nano volts (nV) or micro volts (μV), and the concept of scientific notation of small values is covered in the Chap. 8. However the values of voltage or power are often expressed on a decibel scale.

Consider the input side of a radio (see Fig. 7.3).

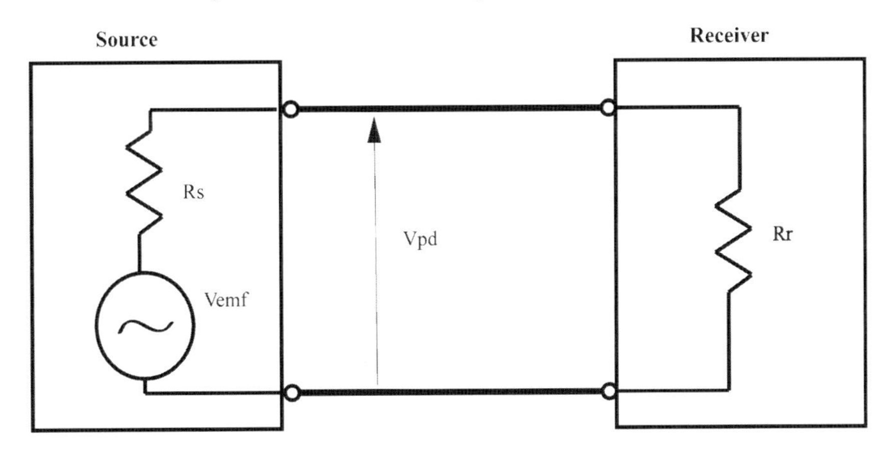

Rs = source resistance

Rr = receiver resistance

Rr should equal Rs for maximum power transfer

Vpd = Vemf/2

Fig. 7.3. The input side of a radio.

The voltage a source can supply if measured in open-circuit mode will give a value of V_{emf}. However, when connected to the receiver, the resistive impedance of the source and radio should match perfectly (for best performance), in which case the voltage appearing across the input terminals V_{pd} is half of V_{emf} because R_s and R_{in} act as a potential divider.

The dBm Unit

The decibel scale is logarithmic and based on the ratio of two values. Since it is dimensionless (it has no units) it needs to be qualified with a small letter appended to the right. Here dBm is decibels relative to one milliwatt (1 mW) as dissipated across a 50 Ω load. Radio circuits are usually designed to have 50 Ω input impedance. The value of dBm is calculated from the following formula. (Note the power ratio is P mW/1 mW, so the 1 mW is omitted from the formula.)

$$dBm = 10LOG(PmW)$$

where LOG is the base 10 logarithm of the power measured in milliwatts so that the power of 0.5 mW is −3.0 dBm and a power of 2 mW is +3.0 dBm.

Note that if the dBm value is less than 0, the power level is simply less than 1 mW, and if positive it is more than 1 mW. Also note that +3 dB on any decibel scale is double the value, and −3 dB is half. This is useful to remember as bandwidth of circuits is defined by their half power points, the difference between the upper and lower frequency where the power has dropped by 3 dB.

The dBmV

As the name suggests the dBmV is the decibel referred to as 1 mV, but this time it is based on a 75 Ω impedance rather than 50 Ω. It is a unit used in defining specifications of television and video equipment, whose characteristic impedance is usually 75 Ω.

The dBμV

The dBμV is the decibel referred to as 1 μV, across a 50 Ω load. The voltage usually refers to the V_{emf}. In order to convert dBμV to dBm simply subtract 113. Therefore 60 dBμV is −53 dBm.

Noise

It is important to consider the noise performance in any radio, but especially important in radio astronomy. Obviously, if the noise generated within the receiver is higher than that of a radio source, the source will not be visible, right? Wrong! The output of a radio receiver is the sum of all the sources of noise. Even if the system noise is high compared to a weak source, the source still adds a little bit to the output. If the receiver is then switched to detect a known calibrated noise source and the difference is taken between the two, then the system noise cancels out, leaving the value of the weak source relative to your calibration source. In this way weak sources can still be detected. This is not a reason to ignore system noise; it should still be minimized as much as possible.

Noise is generated by all electronic components. Careful design of a radio can minimize the effects. It is the temperature of the components that generates thermal noise, and this is most significant in the first stage of a radio, the RF amplifier. Careful selection of low noise components helps a lot in practical designs. Active devices based on gallium arsenide rather than silicon offer good low noise performance, and it is this material that enabled the easy and cheap direct reception of satellite television for home users. The noise performance is good enough to avoid the need for cooling the front end.

The more the front end amplifier can be kept cool the better it will work. It would not be out of the question to thermally insulate the front end and use a Peltier cooler inside to stabilize the temperature. (Peltier cooling is used in many CCD cameras designed for astronomy.) Although a simple Peltier-based system may not reduce the overall noise performance by much, if temperature regulation was used at least the noise performance should be consistent from one season to the next. Cooling systems will not be discussed in these projects, but it would be an interesting exercise to explore the possibilities later, when you gain some experience.

When mounting an RF amplifier consider its placement. If you are observing the Sun with a dish-based antenna, the worst place for the amp is at the focus of a dish. The heat of the Sun will also be focused there and could even damage the amplifier.

Radio receiver noise performance can be specified using one of three possible methods: the noise factor (Fn), a simple ratio value; the noise figure (NF) on a decibel scale; or the noise temperature (Te) in kelvin:

Noise Factor, Fn

The equation for this is:

$$Fn = \left(\frac{P_{no}}{P_{ni}}\right)$$

where F_n is the noise factor, P_{no} is the output noise power, and P_{ni} is the input noise power.

For comparison purposes the F_n value is given for room temperature, often 21°C. In the brackets the ratio is noise power output of the radio, divided by the noise power input.

Noise Figure, NF

The noise figure is simply the noise factor converted to decibels:

$$NF = 10LOG(Fn)$$

The lower the NF the better the receiver is.

Noise Temperature, Te

Noise temperature is a theoretical concept, and it refers to the noise that would be generated by a resistor raised to the temperature Te. The noise temperature is related to the noise factor by:

$$T_e = (F_n - 1)T_o$$

where F_n is the noise factor, and T_0 is the reference temperature in kelvin (standard room temperature is 290 K).

T_e is related to the noise figure by:

$$T_e = KT_0 LOG^{-1}\left(\frac{NF}{10} - 1\right)$$

where K is the Boltzmann constant, T_0 the reference temperature, and NF is the noise factor.

When cascading modules together in a radio system, the significance of the noise performance of each stage down the chain drops very quickly, so the dominating noise is that of the first stage. The combination of noise performance for all the modules in a radio receiver is given by:

$$F_n = F_1 + \left(\frac{F_2 - 1}{G_1}\right) + \left(\frac{F_3 - 1}{G_1 G_2}\right) + \cdots + \left(\frac{F_n - 1}{G_1 G_2 \ldots G_{N+1}}\right)$$

where F_n is the overall noise factor, and the F_1, F_2, etc., are the noise factors of stage 1 and 2, etc., G_1 and G_2 are the gains of stage 1 and 2, etc.

It is clear from this equation that the dominant noise source is the first stage. The contribution of noise presented by subsequent stages quickly drops to a small value.

Sensitivity and Selectivity

Sensitivity refers to how well a radio will respond to weak signals, while selectivity refers to its ability to distinguish between two frequencies close together.

Sensitivity is quoted for radios in micro volts, but it should be referenced by signal to noise ratio in a specified bandwidth. For example an ICOM IC-707 quotes a sensitivity of less than 2.0 μV for a 10 dB signal to noise ratio over the range of 1.8–30 MHz in AM mode. The AM bandwidth for a single channel will be around 6 kHz. If the bandwidth of a channel is reduced, the sensitivity will increase by the square root of the ratio, although wide bandwidths do allow the collection of more power. So there is a tradeoff between received power and sensitivity. If the celestial objects emit radio in a very broad range of frequencies, then increasing the bandwidth will collect more power, and it will be easier to observe the object. However, if the telescope is used to study radio emission lines, broad bandwidth is then undesirable, and a narrow bandwidth will greatly improve the sensitivity.

Selectivity specifications define how great the attenuation is at a given frequency difference. Again the IC-707 states that for AM mode, there is a 6 dB attenuation more than 6 kHz from center frequency, and this increases to 40 dB attenuation at

20 kHz away from center frequency. Note that the specifications will probably show the figures as −6 and −40 dB without mentioning attenuation. The negative sign means loss or attenuation in this case.

Although having a sensitive receiver is important, it is more important to have a good selective receiver. Strong out of band interference is thus reduced or eliminated rather than having a sensitive unit subject to bad noise performance.

Image Rejection

Image frequency is frequency that occurs twice as far away from the radio frequency of interest than the intermediate frequency. Whether that is above or below the channel frequency depends on the local oscillator. In the earlier discussion on mixers it was assumed that the local oscillator had a frequency lower than the radio frequency. It could equally be higher and still provide the same sum and difference outputs. These are referred to as low and high side injection, respectively.

For example, a radio receiver is designed to observe solar flares and Jupiter storms at 22 MHz. The intermediate frequency chosen is 455 kHz because filters are easily obtained for 455 kHz. This means the LO frequency for low side injection is 22–0.455 MHz, or 21.545 MHz.

If there was a strong noise source at a frequency of 2 × 0.455 MHz below the RF frequency, at 21.09 MHz, when mixed with 21.545 MHz, you get an output of 455 kHz and 42.635 MHz. The noise source will pass unimpeded through the system! Note you can't have a negative frequency, so if frequency minus local oscillator is negative then the difference is local oscillator minus frequency

There are a few options open to solve this problem. If you switched to high side injection with an LO of 22.455 MHz the image frequency would now be above the RF channel at 22.91 MHz, which may be clear. Otherwise a different local oscillator frequency must be chosen. Some receivers avoid the problem by using double conversion, first by increasing the RF to a higher first IF, then filtering the required output, and then reducing again to a lower IF.

It is important to understand this concept of image frequency when designing equipment, so go through the numbers above a few times until they are clear.

Third Order Intercept Point, IP3

This refers to intermodulation products (IPs). It's an important parameter, but a bit tricky to explain.

Third order IP products are normally very small when a receiver is operated with normal parameters and will be smaller than the receiver noise floor until the front end is overloaded with a strong signal that saturates the RF amplifier. At this point the amplifier is driven into non-linear operation, generating significant frequency products in a similar way to a mixer. These products then get mixed and enter the signal path as noise.

A given receiver will have a quoted IP3 figure. Refer to Fig. 7.4 for clarification.

Problems occurring where third order intermodulation products are produced are less likely to occur in a radio telescope dealing with weak signals. However to

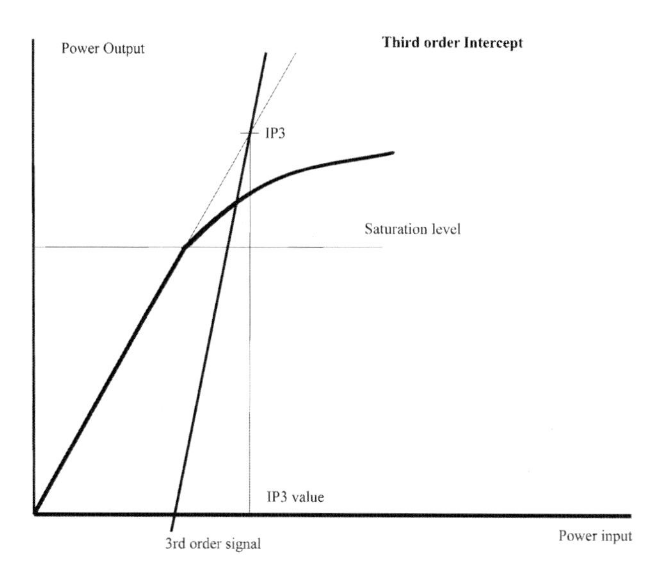

Fig. 7.4. Third order intercept.

stop the problem, an attenuator should be added to the front end, thereby bringing the operation of the front end back into linear operation once again.

For a more in depth description of radio performance the books *Radio Science Observing*, Volumes 1 and 2, by Joseph J. Carr are highly recommended reading.

Chapter 8

Introduction to RF Electronics

This chapter is a brief introduction to electronic components, to circuit building techniques, and to the tools you will need. The aim is to fill in a few gaps for those readers who may well be experienced amateur astronomers but to whom electronics is new. Many texts on amateur radio astronomy have assumed the reader already has skills in radio and electronics. It is not possible in such a small space to cover everything. Indeed, an entire book could be dedicated to this topic alone, so a reading list is included at the end of this book for further information.

The first section of this chapter looks at identifying components, understanding what they can do for us, and learning the symbols used in schematic drawings. More specific examples of using the devices will follow in the project chapters.

The next section discusses circuit building techniques, and the final section describes the essential tools, and some of more specialized but useful tools, for your birthday wish list.

Passive Electronic Components

The definition of a passive component either refers to a component that consumes (but does not produce) energy, or to a component that is incapable of power gain. The opposite is called an active component. So, a resistor that impedes the flow of current is passive, but a transistor which is capable of amplifying signals is active. So let's start with resistors.

The Resistor

See Fig. 8.1 for the possible circuit symbols used for various types of resistor. Note there are two forms, the box and the zigzag. These are interchangeable, but only one style should appear on a given schematic.

Resistor values can be identified by colored bands painted on their body. The majority have four bands, three together and the fourth spaced out. The group of three identify the value, and the fourth band the tolerance of the value. Precision

J. Lashley, *The Radio Sky and How to Observe It*, Astronomers' Observing Guides,
DOI 10.1007/978-1-4419-0883-4_8, © Springer Science+Business Media, LLC 2010

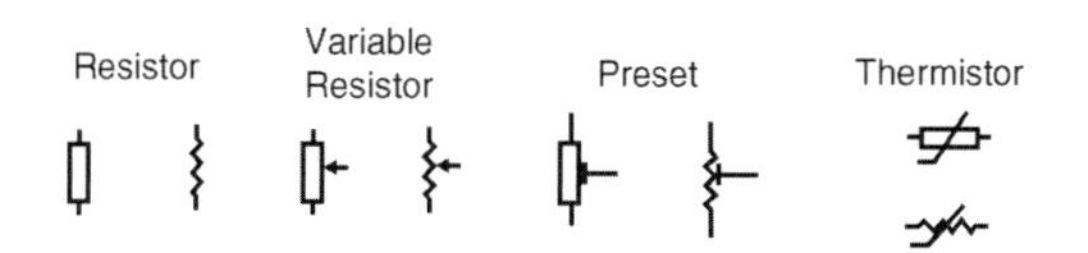

Fig. 8.1. Resistor circuit symbols.

Table 8.1. Resistor color codes

Color	Band 1	Band 2 and 3	Multiplier	Tolerance (%)
Black	0	0	1	
Brown	1	1	10	±1
Red	2	2	100	±2
Orange	3	3	1K	
Yellow	4	4	10K	
Green	5	5	100K	±0.5
Blue	6	6	1M	±0.25
Violet	7	7	10M	±0.1
Gray	8	8		±0.05
White	9	9		
Gold			0.1	±5
Silver			0.01	±10

resistors have five bands, the first four defining the value, and the fifth spaced out the tolerance. The table provides the key to determining their value (Table 8.1).

For example a 56 kΩ resistor with a 5% tolerance would have four bands colored Green, Blue, Orange, and Gold. A 4.7 kΩ 1% resistor would be Yellow, Violet, Red, and Brown. The colors can be surprisingly difficult to read when the body of the resistor is also colored. If in doubt measure it with a multimeter.

Resistors are used to limit the amount of current flowing in a circuit. They can also be used to divide voltages. Firstly consider the circuit in Fig. 8.2.

The important information we need to know about the circuit is how much current will flow. There is a simple formula known as Ohms law which is

$$V = IR$$

or, when transposed:

$$I = \frac{V}{R} \quad and \quad R = \frac{V}{I}$$

The current flow is calculated by dividing the voltage by the resistance, which in this case is:

$$I = \frac{9}{1000} = 0.009A \ (or\ 9mA)$$

Note that when calculating values using this formula, the current is in amperes, voltage in volts and resistance in ohms. However, we often deal with thousands of ohms, or thousandths of amps. Table 8.2 lists the nomenclature used as shorthand to write values that are very large or very small. It is important to understand these they will crop up all over the place.

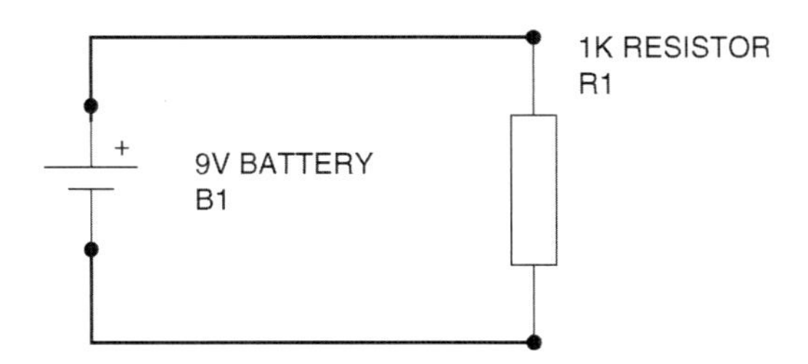

Fig. 8.2. Function of a resistor.

Table 8.2. Dealing with the large and small

Prefix	Scientific notation	Decimal
M – mega	10^6	1,000,000
k – kilo	10^3	1,000
C – centi	10^{-2}	0.01
m – milli	10^{-3}	0.001
μ – micro	10^{-6}	0.000001
n – nano	10^{-9}	0.000000001
p – pico	10^{-12}	0.000000000001

Examples of electronic component values:
10 kΩ = 10,000 Ω
2.7 μF could also appear as 2μ7 = 0.0000027 farads (see capacitors)

The other use for resistors is to split a voltage level. The circuit is known as a potential divider (see Fig. 8.3).

In the example the output voltage can be calculated from the ratio of R2 to the sum of R1 and R2:

$$V_{out} = V_s\left(\frac{R2}{R1+R2}\right)$$

$$V_{out} = 10x\left(\frac{8000}{2000+8000}\right) = 8V$$

For the special case where R1 = R2 the voltage will be half the supply voltage (Vs).

Variable resistors come in two forms, potentiometers and presets. Potentiometers are larger and designed to provide an external control that will need regular adjustment by the user, for example, a volume control. Presets are smaller and usually mounted directly to a circuit board. They are designed to be adjusted infrequently, and possibly only one time, and are used to set up or calibrate a circuit after it is constructed. These devices have three pin and are variable potential dividers. The center pin is the same as the center point in Fig. 8.3.

When combining resistors together in series, their values are simply added together. However, when connected in parallel, the following formula is used (Fig. 8.4):

$$R = \frac{R1R2}{(R1+R2)}$$

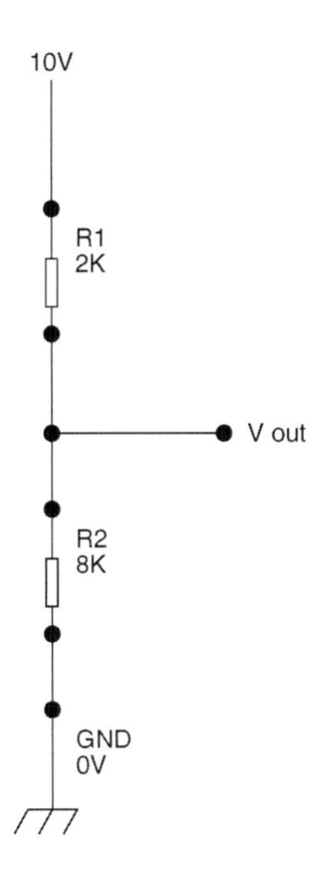

Fig. 8.3. Potential divider.

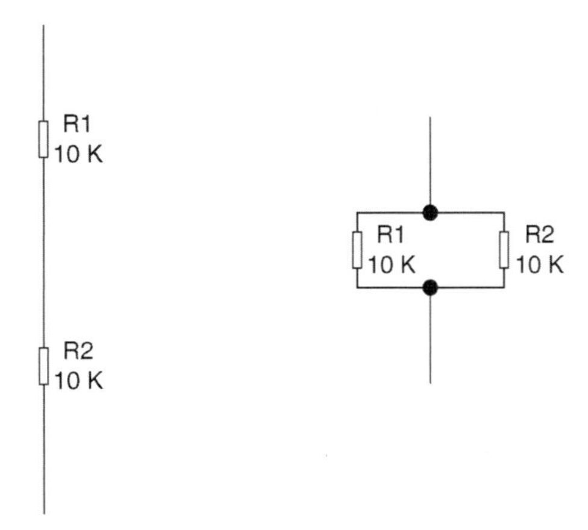

Fig. 8.4. Series and parallel.

Capacitors

Capacitors store energy in the form of an electric charge. In a DC circuit no current will flow through a capacitor, but it will charge up to the supplied voltage across it (Fig. 8.5).

As an experiment take a high value electrolytic capacitor, say about 1,000 μF, with a rating of 15 V or more. This type is polarized, and therefore has a positive

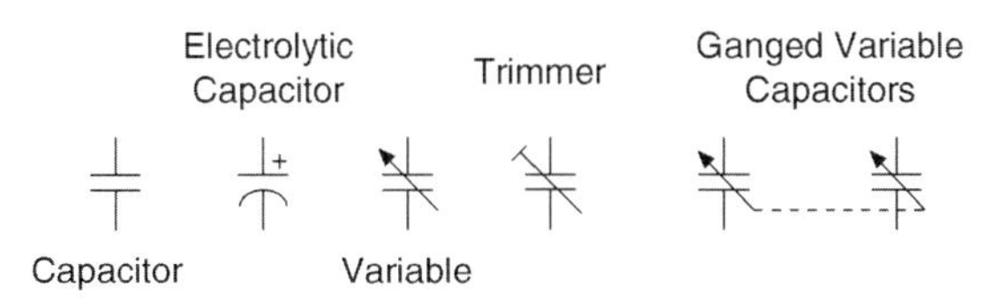

Fig. 8.5. Capacitor circuit symbols.

and negative lead. Connect it to a 9 V battery for a short time, and then as quickly as you can remove the battery and connect a 6 V lamp across the capacitor. Breifly the lamp will light and then fade, showing the capacitor had stored energy. If the capacitor is left for some time in a charged state, energy will slowly leak away due to imperfections. All capacitors leak charge, but some more than others.

In radio circuits it is much more important to understand their performance when subjected to alternating currents, especially sinusoidal radio frequency currents. At radio frequencies a capacitor can act like a resistor does in a DC circuit, that is, it has reactance that is also measured in ohms. Often in radio circuits there are points where it is necessary to block DC currents but allow RF signals to pass through, in which case a value is chosen that has a low reactance at the frequency involved. Devices that behave this way are referred to as bypass or coupling capacitors.

It is important to note here reactance varies with frequency. In general, reactance of capacitors falls linearly with increasing frequency, at least that is the ideal. The real world can be more complex than that! Capacitive reactance is calculated from the following formula:

$$Xc = \frac{1}{2\pi fC}$$

where f is the frequency in hertz and C is the capactitance in farads, and π is a constant of value approximately 3.1415. Xc is measured in ohms (Ω).

Unlike resistors, when capacitors are mounted in parallel, their values are added together. When they are mounted in a series the total value is given by:

$$C_T = \frac{C_1 C_2}{C_1 + C_2}$$

where C_T is the combined capacitance, and C_1 and C_2 are mounted in a series.

There are many types of capacitors available, with their construction and appearance varying widely. They all consist of a pair of metal plates separated by an insulating material known as the dielectric. Their construction varies with dielectric type, capacitance, and voltage handling abilities. Therefore not all types are available in all values. Common types are electrolytic for high values and ceramic disc for small values. There are also various plastic film types. The photographs in Fig. 8.6 show what these look like.

Larger value capacitors in the μF range usually have clear values marked on the body. Small value capacitors often have a three-digit number representing the value in pico farads. So a number 103 means 10 with three zeros, or 10,000 pF, which is equal to 10 nF. The general rule is, if no clear range value is marked (such as nF or μF) then the marked value is in pF.

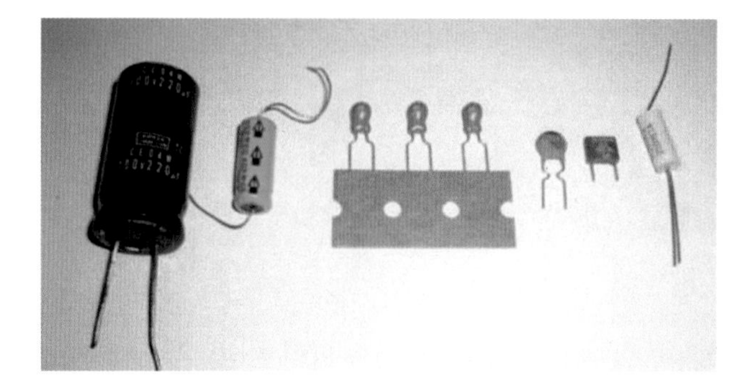

Fig. 8.6. Of all the common components, the capacitor appears in many different physical types. Some of the common ones are shown here. From *left* to *right*: radial electrolytic, axial electrolytic, three Tantalum beads, ceramic disc, plastic film, and high voltage non-polarized electrolytic. Most electrolytic capacitors and Tantalum types are polarized. The size of electrolytic types reflects their values and voltages. They are available in μF values. Disc types are found in pF and low nF values. Film capacitors are mostly available in nF ranges.

Capacitors also come in variable types. Although it used to be common that a variable capacitor was used to tune a radio circuit, such techniques have been largely superceded in modern times, and air-spaced variable models are now difficult to buy new. However, small value trimmer capacitors are still available. They can be used to tune the resonant frequency of a tank circuit (more on this later).

One farad is a very large value. All practical capacitors for our purposes hold very small values, so micro, nano, and pico farads are the values to be seen most often.

Inductors

Inductors are another important component used in radio frequency circuits. They are basically coils of wire wound onto a core, or even free-standing air-spaced if the wire is thick enough to support its own weight. Many types of inductors have a ferrite or powdered iron core that increases the coil's ability to store energy, this time in the form of a magnetic field (Fig. 8.7).

An inductor is the complement to a capacitor. It, too, has a reactance that varies with frequency, only this time the reactance increases with increasing frequency. The formula for inductive reactance is:

$$XL = 2\pi fL$$

where F is again frequency in hertz, $\pi = 3.1415$, and L is the inductance in henries.

Again like resistors, series-mounted inductors add up in value. And the value of two mounted in parallel is given by:

$$L_T = \frac{L_1 L_2}{L_1 + L_2}$$

where L_T is again the combined value, and L_1 and L_2 are parallel connected.

Like capacitance, 1 H is a very large value you will not encounter in practical circuits. Values of micro and nano henries are most commonly used.

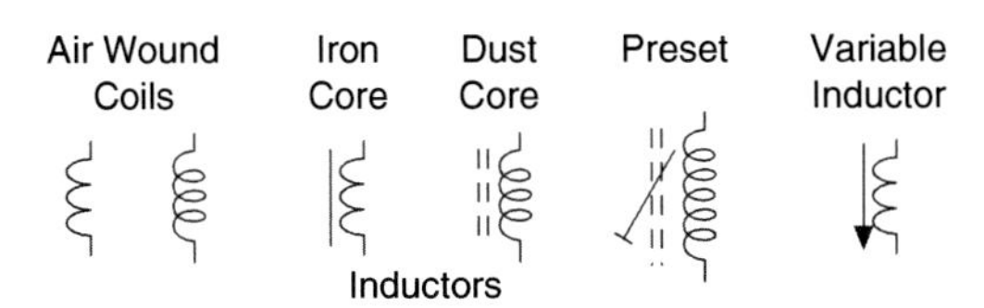

Fig. 8.7. Inductor circuit symbols.

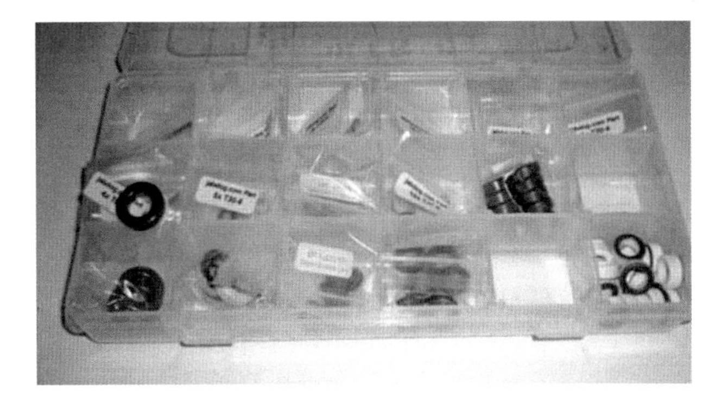

Fig. 8.8. Toroidal inductor cores. Powdered iron types are usually color coded. Ferrites are normally *black*.

Traditional radio circuits employed a number of inductors or transformers usually screened inside a metal can. Valve cicuits employed some physically quite large coils of this type. Once again many of these components are obsolete in modern times. This often makes replicating published designs from the past very difficult.

For the simple circuits covered in this book one type of inductor is very useful, the toroidal inductor. Toroidal inductors can be constructed for most practical values of inductance using a single coil wound onto a ferrite or powdered iron core and can be made into RF transformers by winding a pair of coils onto the same core. Companies such as Micrometals and Amidon can supply a wide range of core types and sizes. It may prove useful to purchase a kit of the more common types (see Fig. 8.8).

The nomenclature used for iron cores follow a pattern such as T-37-2. The T stands for powdered iron toroids, whereas ferrites use the term FT. The middle section, in this case 37, refers to a size in millimeters of the torroid, and the final number, 2, is the mix type. For each core the manufacturer will supply a number referred to as the A_L value. This value varies from core to core and relates to the properties of the mix type and dimension of the torroid. Understanding its derivation is not required, but the following formulae allow the number of turns for a given value for inductance to be calculated.

Ferrite Materials (FT)

$$N = 1000\sqrt{\frac{L}{AL}}$$

where L is in mH (milli henries) and AL is in mH/ 1,000 turns.

Powered Iron Types (T)

$$N = 100\sqrt{\frac{L}{AL}}$$

where L is in μH and AL is in μH/ 100 turns.

In order to effectively use a toroidal core to make an inductor, you must know the mix type, which will tell you the AL value you need. It is pointless purchasing cores if the supplier can't tell you this information (which may be the case for surplus component sales). Tables 8.3–8.5 gives the properties for all types.

Another important form of inductor is the RF transformer. This is used to couple signals between different stages in a radio. Transformer coupling is important where there is a mismatched impedance between stages. It provides a means of transferring the maximum amount of power. In general, you will not be able to purchase readymade RF transformers; there are simply too many variations in their application. The exception is where common intermediate frequencies are used, such as 10.7 MHz. For many applications you will need to construct your own. The traditional way to do it used vertical coil formers with adjustable iron cores, surrounded by screened cans. These are getting more difficult to find now, so the following technique uses toroidal ferrite or powdered iron cores. The only drawback is they are not easily adjustable after construction.

A transformer contains a pair of coils wound onto the same core. The input coil is the primary, and the output coil is the secondary. When constructing use different colored enameled wires for each coil so they can be easily identified when you come to use them. The ratio of the number of turns in the tranformer is related to the ratio of the impedances by:

$$\frac{N_p}{N_s} = \sqrt{\frac{Z_p}{Z_s}}$$

where N_p and N_s are the number of coil turns on the primary and secondary, and Z_p and Z_s are the primary and secondary impedances, respectively.

The general rule in RF transformer design is that the reactance at the lowest frequency must be 4 times the impedance connected to the winding.

Here's an example. The ouput impedance of a tuner module must not exceed 500 Ω but needs to be down converted to a lower frequency by the next stage, which has an input impedance of 1,500 Ω. The IF output of the tuner is a range of frequencies from 33 to 39 MHz. So how do you design a transformer to match the system?

Table 8.3. Powdered iron toroid types

Mix	26	3	15	1	2	7	6	10	12	17	0
Color	Yellow/white	Gray	Red/white	Blue	Red	White	Yellow	Black	Green/white	Blue/yellow	Tan
Frequency (MHz)	DC-1	0.05–0.5	0.10–2	0.5–5	2–30	3–35	10–50	30–100	50–200	40–180	100–300
μ	75	35	25	20	10	9	8	6	4	4	1
Temp coef. (PPM/°)	825	370	190	280	95	30	35	150	170	50	0

Table 8.4. AL Values for powdered iron cores

Mix	26	3	15	1	2	7	6	10	12	17	0
T-12	–	60	50	48	20	18	17	12	7.5	7.5	3
T-16	145	61	55	44	22	–	19	13	8	8	3
T-20	180	76	65	52	27	24	22	16	10	10	3.5
T-25	235	100	85	70	34	29	27	19	12	12	4.5
T-30	325	140	93	85	43	37	36	25	16	16	6
T-37	275	120	90	80	40	32	30	25	15	15	4.9
T-44	360	180	160	105	52	46	42	33	18.5	18.5	6.5
T-50	320	175	135	100	49	43	40	31	18	18	6.4
T-68	420	195	180	115	57	52	47	32	21	21	7.5
T-80	450	180	170	115	55	50	45	32	22	22	8.5
T-94	590	248	200	160	84	–	70	58	32	–	10.6
T-106	900	450	345	325	135	133	116	–	–	–	19
T-130	785	350	250	200	110	103	96	–	–	–	15
T-157	870	420	360	320	140	–	115	–	–	–	–
T-184	1640	720	–	500	240	–	195	–	–	–	–
T-200	895	425	–	250	120	105	100	–	–	–	–

Table 8.5. AL Values for ferrite toroid cores

Type	43	61	63	67	68	72	75	77	F	J
FT-23	188	24.8	7.9	7.8	4	396	990	356	–	–
FT-37	420	55.3	17.7	17.7	8.8	884	2,210	796	–	–
FT-50	523	68	22	22	11	1,100	2,750	990	–	–
FT-50A	570	75	24	24	12	1,200	2,990	1,080	–	–
FT-50B	1140	150	48	48	12	2,400	–	2,160	–	–
FT-82	557	73.3	22.4	22.4	11.7	1,170	3,020	1,060	–	3,020
FT-87A	–	–	–	–	–	–	–	–	3,700	6,040
FT-114	603	79.3	25.4	25.4	–	1,270	3,170	1,140	1,902	3,170
FT-114A	–	146	–	–	–	2,340	–	–	–	–
FT-140	952	140	45	45	–	2,250	6,736	2,340	–	6,736
FT-150	–	–	–	–	–	–	–	–	2,640	4,400
FT-150A	–	–	–	–	–	–	–	–	5,020	8,370
FT-193A	–	–	–	–	–	–	–	–	4,460	7,435
FT-240	1240	173	53	53	–	3,130	6,845	3,130	–	6,845

The turns ratio of the transformer is given by:

$$\sqrt{\frac{500}{1500}} = 0.577$$

We require an reactance on the secondary side of 1,500 Ω × 4 = 6,000 Ω. The inductance in μH of the secondary winding can be found from:

$$L = \frac{6000\Omega}{2\pi f}10^{6}$$

$$\mathrm{L} = 28.9\mu\mathrm{H}$$

Using the earlier formula for calculating toroid inductance and choosing the ferrite core FT37-43 with an AL of 420:

$$L = 0.0289\text{mH}$$

$$N = 1000\sqrt{\frac{0.0289}{420}}$$

$$N_s \approx 8 \text{ turns}$$

Therefore

$$N_p = 8 \text{ x } 0.577 \approx 5 \text{ turns.}$$

Note here how to determine the number of turns in a toroid core. It is the number of times the wire passes through the center. If, for example, the wire was passed through wrapped around the core and passed through again this is two turns, not one.

Crystals and Resonators

Crystals and ceramic resonators are very important to radio electronics in order to control the frequency of oscillators, for example.

Crystals operate by what is known as the piezo electric effect. If a oscillating signal is applied to the crystal, the device is induced into vibrating mechanically. Equally, if the crystal is mechanically vibrated, it will generate electricity (Fig. 8.9).

The first property is the most important to us in RF engineering. The crystal itself does not generate an oscillating electrical waveform; rather, it is used to regulate an existing oscillator and keep it accurately on frequency.

It is important to understand that all crystals are constructed to have a fundamental mode of vibration at a given frequency. Fundamental refers to the lowest frequency it will operate at. However, crystals will also oscillate at higher frequencies that are approximate multiples of the fundamental. These higher frequencies are referred to as overtone frequencies. Note that overtones are not exact multiples of the fundamental frequency. The term harmonic is used to mean exact multiples of a fundamental frequency.

In the construction of crystals, physically making them thin enough to vibrate at high frequencies becomes impossible for frequencies above about 20 MHz.

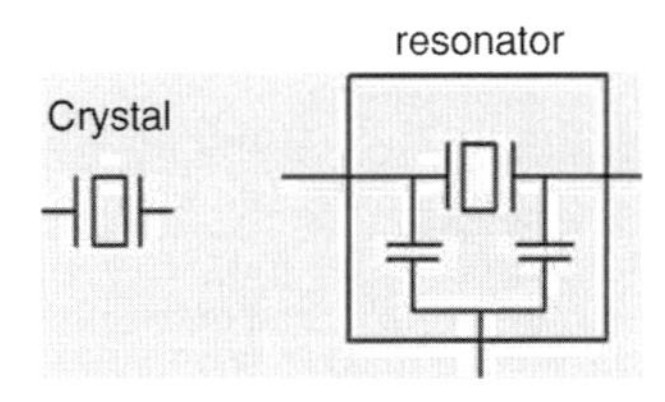

Fig. 8.9. Crystals and resonators

Higher frequency devices are known then as overtone crystals. The frequency marked on the crystal body is the frequency it was designed to operate on, but not necessarily the fundamental frequency. For example, a 27,000 MHz device will probably have a fundamental frequency of close to but not equal to 9 MHz. The device is therefore a third overtone crystal and was manufactured to work at precisely 27,000 MHz, the three zeroes indicating it is accurate within 1 kHz at least, whereas a crystal marked 9.4486 MHz is likely to be operating at its fundamental frequency.

Due to the properties of overtones, it is not recommended to use a crystal at different frequencies other than what is marked on the body. If only a number appears, the markings on the crystals will be in kHz, but high frequency crystals will more than likely show MHz after the number.

A crystal has two pins, but resonantors have three pins. Resonators can be crystal based or may be constructed of piezo electric ceramics. The reason for having three pins is because they contain a pair of internal capactitors. Resonators are usually used as filters, where the center pin is grounded and a signal applied between pin 1 and ground, and the filtered output taken from ground and pin 3. Resonators are not as accurate as crystals for controlling frequency and therefore are not suitable for making precision oscillators.

Diodes

Diodes are the simplest of the semiconductor devices. Most diodes are used to allow current to flow in one direction only, although there are some useful variants available.

Diodes may be found in power supplies, where they are used to rectify AC currents as a means to convert them to DC. They are also found in many radio detector and mixer circuits.

Silicon diodes are the most common general-pupose ones. In operation they do have some forward resistance (and of course a very high reverse resistance), so there is a small voltage drop across them. The voltage drop is typically 0.6 V for silicon devices but it can be as low as 0.15 V for schottky diodes (Fig. 8.10).

Special forms of diode include the varactor and the zenner. Varactors are effectively used in reverse. The function of the PN junction in reverse creates a region inside the diode that is an insulator preventing the flow of DC current, but it acts as a capacitor to AC current. By varying the DC voltage across the varactor, the insulating region will grow or reduce, thus varying the effective capacitance. Varactors can be used to control the frequency in variable oscillators.

The zenner diodes are used to provide a constant voltage to stabilize a power source. They are marked with the voltage at which they will operate.

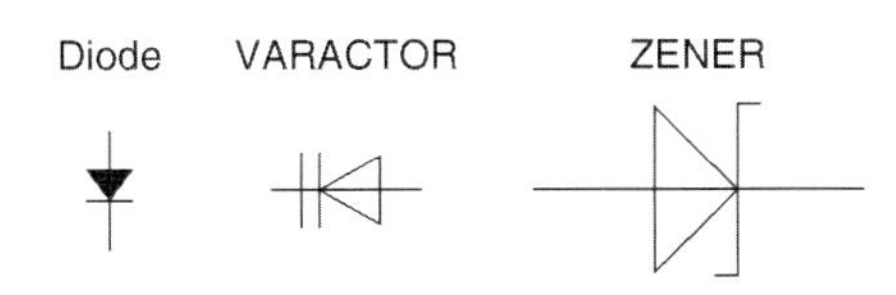

Fig. 8.10. Diode circuit symbols.

Active Components

Bipolar Transistors

The bipolar transistor is an example of an active component. So far we have only looked at passive components. An active component can amplify in some way – the output power is greater than that applied to it.

The BJT is the basis of most circuits. The modern IC chip is simply a miniaturized circuit of transistors and other components. BJT devices are available in two types, NPN and PNP. This refers to their construction, consisting of three layers of semiconductor, usually silicon these days. The N refers to the silicon, which is doped with an impurity that provides it with extra free electrons. P-type material is doped in such a way that the number of electrons are reduced, creating gaps in the crystalline structure known as holes. It is beyond the scope of this book to explain how a transistor functions internally, but there are many texts available that do this. It is only important that you can apply transistors to practical circuits. There are many instances in electronics where a chip or a readymade module is used effectively without the constructor knowing the internal workings of the devices. Not only is a transistor capable of amplifying signals, it can be useful as a solid state switch, too.

The symbols for transistors are shown in Fig. 8.11. The *arrows* in the symbols shows the direction of conventional current flow. Amplifying ability is achieved by applying a small current to the base pin, which allows a much greater current to flow across the collector and emitter pins, typically 50–250 times as much.

Although all transistors function in the same way, the selection of which type to use in a new design is largely based on the operating frequency and the power handling ability. The manufacturer describes in the datasheet what the design role is. For example, high current types may be for power amplifiers or power switching, while low power devices are more suited to small signal amplification and high speed switching in oscillators. While we are discussing datasheets, we should mention that the Internet is a very useful resource for information on active devices. Several websites are dedicated to offering free searchable access to them.

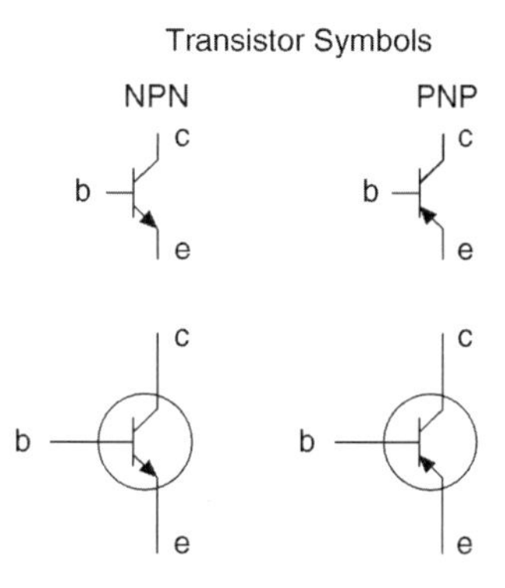

Fig. 8.11. Transistor circuit symbols.

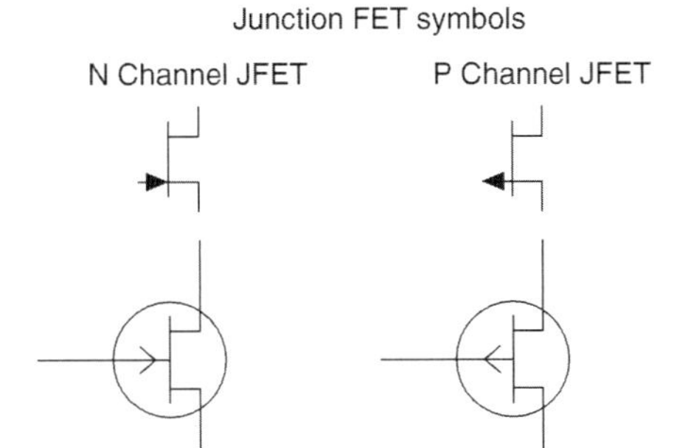

Fig. 8.12. Junction Field Effect Transistor symbols.

Field Effect Transistors (FETs)

The FET is similar to the transistor, in that it is also capable of amplifying signals, but this is a voltage-controlled device. Small devices are known as Junction Field Effect transistors. High power devices are constructed a little differently and are usually called MOSFETs (Metal Oxide Silicon Field Effect Transistors).

Junction FETs need to be handled with care. They have extremely high input impedance, which makes them sensitive to static electricity that can destroy them. Many IC chips that use FET inputs are similarly sensitive. It is therefore recommended to take precautions. MOSFETs are usually constructed with internal protection diodes, which make them easier to handle, but it is still recommended to use a static safe workbench.

Like BJTs they come in two types, N channel and P channel. However, P channel are very rare in JFET types. For the puposes of these projects only N channel devices will be considered (Fig. 8.12).

Power Supplies

Always use regulated power supplies in radio circuits. These can be purchased cheaply readymade. However, it is sometimes necessary to guarantee that a DC supply is stable, so you will also find a voltage regulator chip helpful. It is also useful in situations when more than one voltage is needed in a circuit. For example, the main circuits may have been designed to work on 12 V DC, but a particular IC chip will only work at 5 V. A 5 V regulator can be used to convert 12–5 at the point required.

Voltage regulators can be easily obtained for common voltages such as 6, 9, and 12 V and can handle currents up to 1.5 A in some cases. They are three-pin devices and usually only require a couple of capacitors to function at their best.

The circuit in Fig. 8.13 can supply up to 1.5 A. If a fixed value regulator is used, omit R1 and R2, and the center pin is connected to ground. The input voltage may be lower than 28 V, but it should be at least 3 V higher than the maximum output

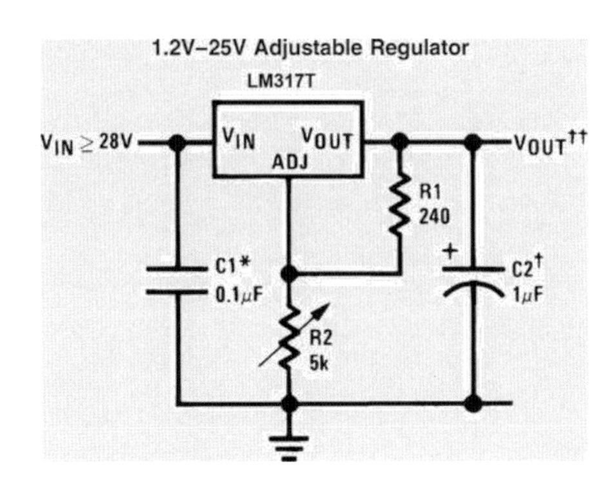

Fig. 8.13. Adjustable voltage regulator application circuit.

required. Note the application as displayed is used only to stabilize an existing DC power source. It is beyond the scope of this book to cover construction of mains AC/DC converters.

AC and Radio Frequency Signals

Before discussing some important information on how radio signals are affected by RF circuits, lets look at what a radio frequency of RF signal is.

A radio signal in its most pure form consists of a time-varying electric component and a time-varying magnetic component with sinusoidal profiles, which are at 90° to each other (Fig. 8.14).

However, for the most part in this book when we talk about an RF signal in a circuit it is in fact an AC electrical signal that was produced in the antenna and has the same frequency as the radio wave that was collected.

Real-world radio signals are made up from a collection of different frequency sinusoidal components, and all radio receivers collect at least a small range of them (narrow bandwidth). Radio astronomy receivers usually have much larger bandwidths than commonly found in communications receivers.

For the most part receivers are designed as though they are only receiving one pure sine wave signal, which is chosen to be the center frequency of a small range of interest. The bandwidth can be then designed to suit its function.

It should be noted at this point that it is not necessarily as easy as you think to measure the value of an RF signal. What value? Refer to Fig. 8.15. The parameters we need to know are amplitude (or peak to peak amplitude) and the frequency. The frequency determines how many full cycles occur in a second and is the reciprocal of the time it takes for one cycle to occur, i.e., $f = 1/t$.

When measuring the amplitude of an AC signal, you can't always rely on the AC range of a multimeter. Most multimeters are designed to test AC power circuits that have a low frequency of 50–60 Hz. They will prove inaccurate, and probably extremely inaccurate when the frequency is many megahertz. True, RMS meters

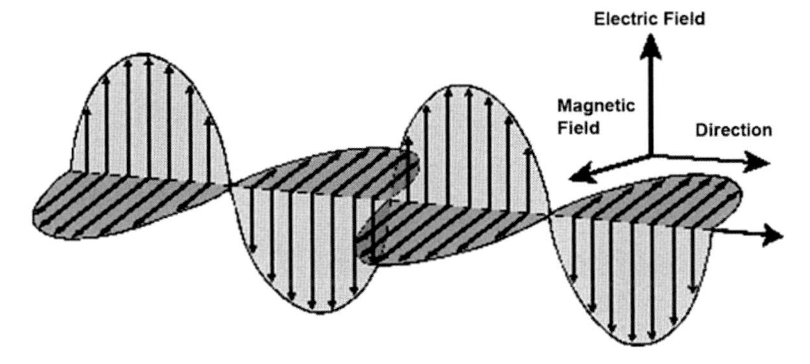

Fig. 8.14. Electromagnetic wave.

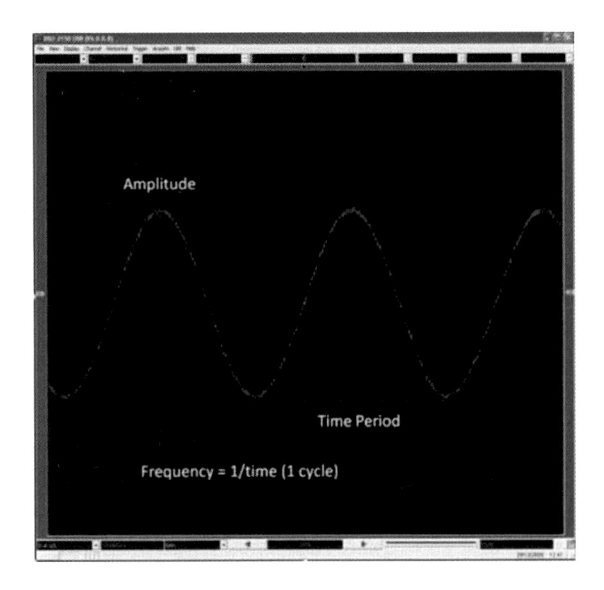

Fig. 8.15. Amplitude and frequency of a sine wave.

will work, but they are still likely to have an upper limit in their frequency response, which is provided in their specifications. Specialist RF voltmeters can be obtained, but these are rarer now than they used to be.

An oscilloscope (see later) can be used to view the signals on a screen, but most low cost units will only work well up to 20, 60, or even 100 MHz, which are useful but fall short in performance for many of the signals we deal with.

The best tool available for analyzing radio signals is the spectrum analyzer, which presents a graph of amplitude against frequency, rather than amplitude against time. High frequency units, though, are very expensive.

Finally, when we talk about the voltage of an AC or RF signal, it needs to be made clear what this represents. It is not the peak value. The voltage of an RF signal is known as the RMS, or root mean square value. It is a sort of an average value.

The RMS voltage can be found from the peak voltage (as seen on an oscilloscope screen) from:

$$V_{rms} = \frac{V_{peak}}{\sqrt{2}}$$

The Tuned (Tank) Circuit

The tuned circuit is found in all radios and is a combination of an inductor and a capacitor. It is the basis of defining the frequency of response of an RF circuit. There are two forms, series connected and parallel connected. It is important to understand the differences.

Series Resonance

Consider the circuit in Fig. 8.16. The upper circuit is an example of series resonance, and the lower circuit parallel resonance.

The voltage in the inductor will lead the current flowing through it by 90°, and the voltage across the capacitor will lag the current by 90°. This means the two components oppose each other and tend to cancel out. However, the voltages across the components depend on their reactance. Reactance is a form of resistance but only applies to AC signals, not DC. As we saw earlier, reactance is calculated from the following:

$$X_L = 2\pi fL$$

$$X_C = \frac{1}{2\pi fC}$$

where X_L is the inductor reactance, X_C is the capacitor reactance measured in ohms, L in henries, and C in farads. The frequency in hertz is f.
From Ohms law:

$$V_L = IX_L$$

$$V_C = IX_C$$

where V_L and V_C are the voltages across the inductor and capacitor, respectively.

At a particular frequency, the reactance of the inductor and capacitor will be equal, and the voltages will exactly cancel out. This means:

$$2\pi fL = \frac{1}{2\pi fC}$$

Rearranging the formula we can solve it for f:

$$f = \frac{1}{2\pi\sqrt{LC}}$$

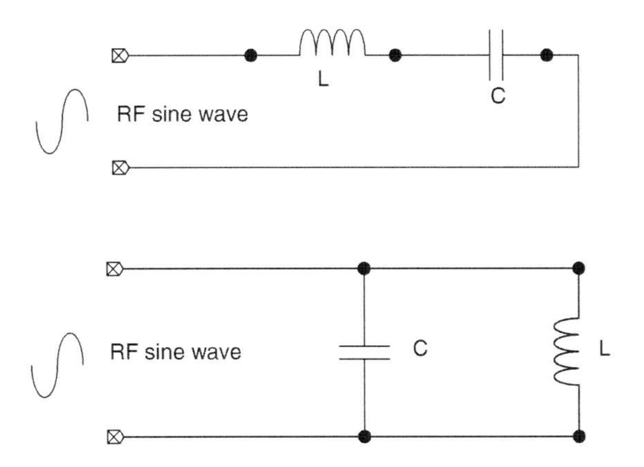

Fig. 8.16. Resonant circuits.

Although it is not necessary to remember the derivation of this formula, the result is extremely useful when designing radio circuits. In order to determine the values of L and C required for a given frequency, usually the value of C is preselected, and L is calculated. This way it is possible to build your own inductor to resonate at any reasonable frequency by using readily available capacitor values. When either L or C is preselected the other can be calculated eaily from:

$$C = \frac{1}{39.5 f^2 L}$$

or

$$L = \frac{1}{39.5 f^2 C}$$

where L is in henries, C is in farads, and f is in hertz.

The resulting values of L and C are very small numbers and need to be converted (usually) into micro henries and pico farads.

The important consequence of this for a series resonant circuit is that the impedance is at a minimum value at the point of resonance. Impedance will be theoretically zero, but in practice a small amount of pure resistance will be present, so a small loss is possible. Therefore a series tuned circuit will allow resonant frequency signals to pass through, but will impede (or reject) unwanted frequencies. This is sometimes known as an "acceptor" circuit. A practical use of this is found in chapter 12, see Fig. 12.3 for an oscillator circuit using series resonant components.

Parallel Resonance

In the case of parallel resonance it is the currents in the components that cancel out at resonance. The same formula applies to calulate the frequency at which this occurs, but now the impedance is at a maximum. This is a "rejector" circuit and will block the flow of RF signals at the resonant value.

Q Factor

The Q factor is a numerical representation of how sharply tuned a resonant circuit is. High Q is a sharp, small bandwidth response. The definition of Q is the ratio of the reactance to the pure resistance of the circuit. The resistance is mostly provided by the inductor. When an inductor and capacitor are used in a tuned circuit such as parallel tank circuit, the impedance at resonance is not infinite, due to the imperfections in the components, mostly the coil resistance. The resistance in this case is usually called dynamic resistance, or sometimes dynamic impedance, and is given by:

$$R_D = \frac{L}{CR}$$

where L is in henries, C in farads, and R in ohms. R_D is the effective resistance to rf signals in ohms.

The Q can then be found from:

$$Q = 2\pi fCR_D$$

or

$$Q = \frac{1}{2\pi fCR}$$

or

$$Q = \frac{2\pi fL}{R}$$

Q can also be found by dividing the center frequency f by the half power bandwidth. Q then is simply a figure of merit, or quality factor, of a tuned circuit.

Coupling Decoupling and Blocking

In radio circuits it is necessary to pass RF signals from one stage to another, but it is usually important to isolate DC volatge conditions between the two stages. The capacitor is used for this purpose. DC voltages are blocked by the capacitor, but so long as the reactance of it is low at the frequency involved the RF signal will pass through virtually unaffected. This is known as coupling. The values of capacitor for RF signals are usually a few pF up to about 0.01 μF.

Decoupling is essentially the opposite. It is the process of removing an RF signal and allowing the DC to pass. This can be done in two ways. A low reactance capacitor can be connected from a circuit point to Earth. RF signals will prefer to take the low impedance path to Earth, and very little will continue along the circuit path. The other way is to fit an inductor in series with the circuit. The value of L is chosen to be a high reactance at the operating frequency, therefore absorbing the RF signal but presenting a very low resistance to DC currents. The inductor approach is often used when feeding power into an RF feeder in order to power remote amplifiers. The capacitor technique is often used at points in an RF circuit.

Transistors as Amplifiers

The circuit in the diagram is the general form of an transistor amplifier using an NPN transistor (Fig. 8.17).

Consider what happens when a small AC or RF current is applied to the input. As the current rises, the current flowing through the transistor from collector to emitter rises, too. Ignoring the presence of C3 for the moment, R3 and R4 form a potential divider. The amount of voltage dropped across R3 increases, so the collector voltage falls. As the input current drops again and turns negative the current flowing decreases, and the voltage at the collector increases. Therefore the output is phase-shifted 90° with respect to the input.

The input current only has to be small to turn on the transistor, but the current flowing through it will be defined by R3 and R4 and can be many times the input current so amplification takes place.

The configuration in the diagram is known as the common emitter, since the emitter pin is common to both the input and output. The transistor will present about 1K of input impedance, and 5K output impedance, but the values of the resistors in the circuit will modify that. Note that sometimes R2 is ommitted, and R1 is connected between the collector and the base. Also, sometimes R4 is omitted.

There are other forms of layout known as the common collector or common base and the emitter follower where the output is taken from the emitter pin rather than the collector. For a more in-depth discussion of amplifiers see the reading list at the end of this book.

In order to function correctly, the transistor needs to be biased. This is where the DC voltage conditions are set up to enable the transistor to work within its dynamic range. If the DC voltage at the base pin is too low, then the positive peaks will "clip" or saturate at the positive supply voltage. If the biasing is too high, then the signal will clip at the minima and distortion also occurs.

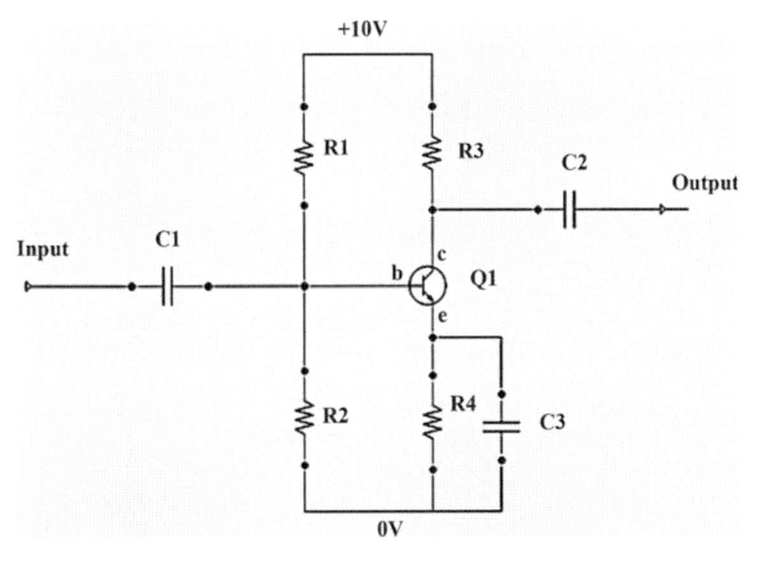

Fig. 8.17. Basic common emitter amplifier.

The resistors R1 and R2 form a fixed potential divider setting up the DC voltage at the base for best operation. R4 is usually small compared with R3, and a typical emitter voltage would be 1.5 V. The voltage at the collector needs in general to be centrally placed between the 10 V supply and ground at, say 5, volts to allow the output to swing either way without distortion.

If the design current through the transistor is to be 1 mA then R4 is 1.5 V/1 mA or 1,500 Ω and R3 = 5 V/1 mA or 5,000 Ω.

There is a voltage drop across a PN junction, i.e., between emitter and base of 0.6 V, so that would place the base voltage at 2.1 V. The actual current drawn by the transistor will be about 10 μA, but to avoid the base current variations upsetting the voltages, we will allow for 100 μA. Therefore from ohms law, R1 = (10 − 2.1 V)/100 μA = 79 kΩ and R2 = 2.1/100 μA or 21 kΩ.

Capacitors C1 and C2 block the DC voltages from affecting the operation of circuits either side, but allow RF signals to pass; they are therefore coupling capacitors. C3 is a bypass capacitor and is chosen because it has a low reactance at the operating frequency. Without C3 the gain of the amplifier will be lower, as R4 will reduce both DC and RF currents.

Circuit Construction

When you think of an electronic circuit you probably picture the PCB, or printed circuit board, used in most modern commercially built instruments. You may think, how am I ever going to be able to build something like that? The answer is, you may not have to.

Constructing simple PCB's is not that hard. You start with a blank board that will have copper deposited on one side, or maybe on both sides. If the circuit has only a few components, and small size is not important, you can draw the layout with a special etch resist pen purchased from an electronics supplier. Even nail varnish will work if you cannot get an etch resist pen quickly enough.

Once you are happy with the layout and the copper is adequately protected where you want tracks to exist, the board is submerged in a bath of ferric chloride. Ferric chloride is purchased, as mentioned earlier, from electronic suppliers as a dry powder and mixed with water according to the instructions. It will keep for a long time in a storage jar. One thing, however, is that you need to be careful with it, as it will stain clothes, fingers, and lots of other things badly and should not come into contact with skin or eyes. So wear protective gloves and glasses. The fluid will dissolve bare copper on contact but will take 5 or 10 min, so just let it work, but don't leave it too long, as it will undercut the resist eventually. Once completed wash down the board with clean water and dry it. The resist can be removed with a solvent such as acetone. Once again protect skin and eyes from the acetone, and don't let it come into contact with any plastics, as it may well dissolve them!

Where more complicated circuits are needed, especially those containing IC's, a different technique is required. Where one-off or small hobby production is concerned, you can buy transfer sheets that are designed to take the print from a laser printer. Laser printer toner is a thermoplastic and melts when heated. It acts as a kind of glue that sticks to the copper. The printed sheet is placed face down on the copper and heated with a domestic clothes iron. When cool, peel off to reveal the protected tracks of your design. Check to make sure there are no missing bits, and touch it up with a resist pen if there are.

For simple single-sided boards where the copper tracks are on the bottom, the print is not mirror imaged, but when transferred to the copper it is reversed when you look at the copper side. The board is best drilled first and then etched as before. Special small electric drills can be purchased for the task and small drill bits obtained from good electronics stores. It is best to use a drill stand, or you will forever be breaking off the tiny bits that are typically less than 1 mm diameter. Figure 8.18 shows the transfer process.

Another popular method for PCB manufacture is to have it made for you by specialist manufacturers. The cost of one-off's is relatively high. You need to prepare the artwork for the layout and send it off; a few days later you get your board. The cost falls with quantity, so it would be an idea to use these services if a club were to build several circuits of the same type for members.

For small prototype and one-off construction it is easier to use the dead bug style of construction, also known as ugly style. Dead bug technique works remarkably well for the kind of radio frequency circuits we will deal with in these projects. Figure 12.4 shows a picture of a mixer-built ugly style.

Dead bug refers to the components being mounted often upside down onto a copper-clad board with their legs in the air. Metal can-style transistors can be soldered directly to a copper board; plastic-bodied variants will need a dab of glue to hold them. The copper board is not etched at all this time, because the copper surface acts as a grounding plane, which helps to keep high frequency circuits stable. After all an oscillator, which is inherently unstable by design, is simply an amplifier with a feedback loop. If high frequency amplifiers are badly constructed, accidental feedback through poor grounding could make them oscillate nicely. This is one reason why strip board is not suitable for RF circuits. The longitudinal copper strips can act like capacitors or inductors at high radio frequencies. Also prototyping breadboard is useless for testing high frequency RF boards. This author tried building a 27 MHz oscillator on breadboard once, but it really wanted to oscillate at 87 MHz and there was nothing that could be done about it!

Most metal can transistors have one of the legs attached to the case internally for NPN types; this is usually the collector. It may well be necessary to attach an "island" first. Cut small pieces of double-sided copper board, and solder one face onto the main board. The upper surface of the island can then be used to solder items such as a transistor, or the positive voltage feed. Once you have mounted the main items, the pins can then support interconnecting components such as resistors, capacitors, etc. Try to keep the construction neat, folding component legs with small pliers. In many cases it is convenient to lay out the circuit as it appears in the circuit diagram to make it easier to understand later. The photograph of an RF mixer shows a completed working converter used by the author for testing.

When constructing RF circuits it is often best to separate circuits into modules, built on different PCBs, and possibly housed in separate, metal screened boxes, the interconnections being made by coaxial patch cables for the signals. A complete radio may be broken down into maybe four or five modules. By making each module independently you can test them as you go, making sure each works as it should, so the final design should operate as expected when all hooked up. This method also makes it easy to convert the receiver later to, say, operate on a different frequency just by changing one or two modules, or by making multiband radios by switching in alternative "front ends."

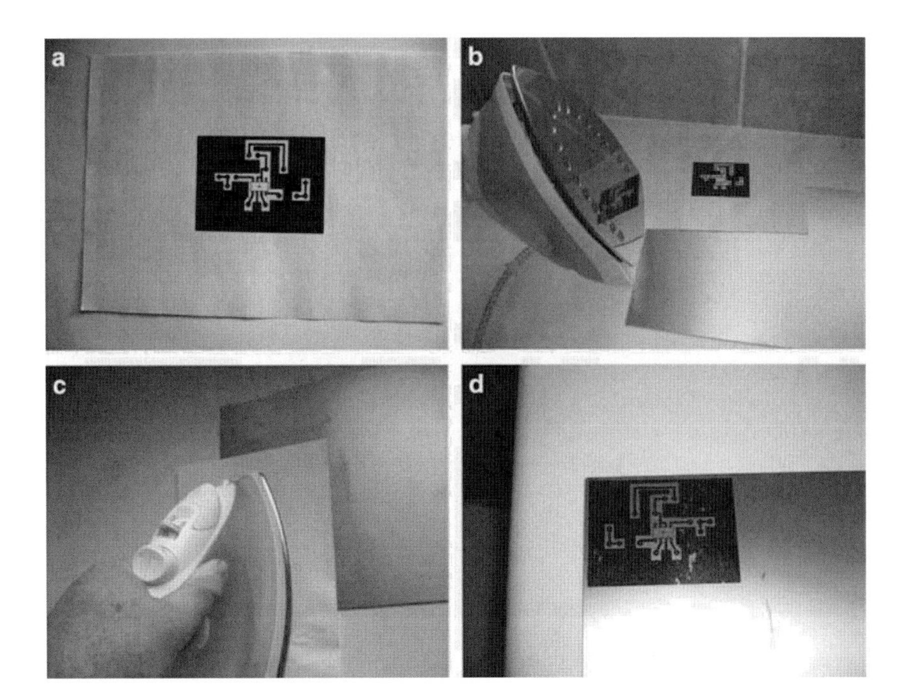

Fig. 8.18. (**a–d**) Using a laser-printed transfer for PCB construction. Note that when the image was transferred to the copper lots of missing patches existed. These were touched up with lacquer from an etch resist pen. None of the important tracks were damaged. Also, an outline of the IC had been left in the print by mistake. This was scratched off with a fine screwdriver before etching.

Screening

Careful thought needs to go into the layout of RF circuits. It is important to isolate stages from each other to avoid RF coupling interfering with the operation of the circuits. For example, the magnetic field from a tank circuit may be coupled through free air to a neighboring element of the circuit. An oscillator will radiate a low power RF signal, which may then be picked up elswhere in the circuit, causing self interference.

Screening is the process whereby possible sources of RF coupling are blocked by mounting them inside metal enclosures that are earthed. Sometimes all it takes is to fit a metal wall between two sections, which could be a piece of copper clad circuit board soldered to a grounded part of the circuit. Cylindrical coils may need screening in a can, but toroidal coils are self screening, which means the field is contained within the iron ring, and only a small separation is needed to avoid problems.

Test Instruments and Tools

Basic Tool Kit

The beginner has to start somewhere. You will probably build up an extensive tool kit in time if the bug really bites. To start off with the following list is essential:

- SOLDERING IRON AND STAND. Temperature control workstation soldering irons are available at fairly low cost and have the advantage of a good weighted stand/ power supply. Small hand held ones are a little cheaper. For PCB work a pointed tip rather than chisel tip unit is best. If the heat is not variable, then a 15–20 W iron is better than high power ones for electronics.
- SIDE CUTTERS. The smallest side cutters are best for close cropping component legs, but a larger pair is useful for cutting wires. Don't abuse side cutters by trying to snip hard metal items or even circuit boards – you may well break them!
- SMALL SCREW DRIVERS, a mixed set of flat and cross head types.
- NEEDLE NOSE PLIERS, with or without cutting blades.
- MAGNIFIER, used mostly as a head shield magnifier, especially if you intend to work with surface mount components.
- MULTIMETER. Most multimeters are digital these days, which is certainly recommended if you only have one. The cheapest ones are not good value, because they have limited ranges and are not as accurate as you may think. It is worth investing in a good brand and model. Research the market, or ask your local amateur radio society for recommendations. The basic ranges are DC and AC volts and amps, resistance, and diode test. A capacitance and frequency range is useful, but it would be better to get a quality meter without these than a poor one with them.
- DRILL AND DRILL BITS. Useful when constructing antennae, mounting components in boxes, etc. A PCB drill and drill stand will be needed if you choose to etch your own circuit boards.
- HACK SAW, not only for cutting the obvious things but for cutting small circuit board pieces, too.

In order to properly test electronic circuits you build, the following items are really useful to have. Although you may be able to get by with the basic kit at first, you will need some or all of these instruments at some point. By joining a local amateur radio club you may be able to get access to these tools at club meetings or even borrow from your friends until you can source your own.

- RF SIGNAL GENERATOR. The RF generator is used to simulate real radio signals when testing RF amplifiers, mixers, etc. RF signal generators can be very expensive in wide band high frequency models, even on the surplus market, but they are always useful to have and should be part of your test bench.
- OSCILLOSCOPE. These days it is easy to pick up a useful working oscilloscope on the surplus market reasonably cheaply. Oscilloscopes are very useful at visualizing what is going on inside circuits. Low-cost units can be purchased from eBay to connect to a PC via USB. The computer then acts as the display. Often they have a bonus feature of a spectrum analyzer, too. Their advantage is the ability to store traces on the PC.
- FREQUENCY COUNTER. A digital frequency counter is handy to confirm the operation of oscillators, or the output frequencies of mixers when used with care. Surplus units that operate to UHF frequencies can be found at reasonable cost. However, as with all surplus or used test equipment, their accuracy can suffer with age. It is not worth buying an old tired inaccurate instrument for a low price and then spending hard-earned money having it calibrated or repaired. Buy the best one you can up front – better if it comes with a dealer's warranty if you are not sure of its quality.

- Step attenuator, which can be bought or made at home, as we shall see. Step attenuators are useful in controlling the output of test oscillators, or the noise source we will construct.
- Noise source. Good calibrated noise sources can be expensive, but later we will see how you can make one. A noise source is handy to simulate the kind of signal we see from space, as well as in testing the performance of amplifiers.
- Impedance bridge. Impedance bridges were once a common tool in the workshop and have been largely superceded by digital instrumentation. However, they can be really useful in testing antenna impedances. Home construction is simple.

Finally, in any RF engineer's wish list is the following instrument:

- Spectrum analyzer. This is a great tool for testing radio circuits. An oscilloscope measures how a signal varies with time; a spectrum analyzer measures how signals vary with frequency. Spectrum analyzers can help you sort out the function of filters, mixers, RF amplifiers, and do many other tasks. However, good ones are seriously expensive new, and even pretty expensive used. However there are some useful if limited performance options available. There are a number of USB computer-based oscilloscopes on the market with useful bandwidths. As a bonus with certain models you can get spectrum analysis, too, which use FFT mathematics to generate the spectrum. The bandwidths are limited on the low-cost models to around 250 MHz or less, but that is still useful. A spectrum analyzer can be obtained for setting up television systems. It has many limitations, but it is still useful for examining the output of RF circuits in the range of 45–2,150 MHz. This can be augmented with a USB oscilloscope/analyzer to fill the gap below 45 MHz.

Most of the tools listed here should be purchased either new or from a surplus source. Online auctions can be a useful source of used test equipment, but it is a minefield out there, and the equipment functions should be established first, unless you need lots of interesting door stops.

Three useful tools are easily constructed that will prove very useful in setting up a radio telescope. The step attenuator, the noise source, and the impedance bridge.

Building the Impedance Bridge

The bridge circuit is a valuable tool to use in testing antenna impedance. It will take out a lot of the guesswork when setting up. Where an antenna has some adjustment, such as the gamma match, or where we make up a one quarter wavelength matching section, the bridge will allow us to tune it correctly.

To understand how a bridge works first consider the circuit in Fig. 8.19. We have seen a similar circuit earlier, the potential divider. This circuit is a pair of potential dividers. If R1 to R4 resistors are all of equal value, say 50 Ω, then the voltage across the output will be half the input voltage, in this case 6 V. Let's now keep R1 and R2 at 50 Ω but make R3 and R4 75 Ω; the output voltage is still half of the input, or 6 V.

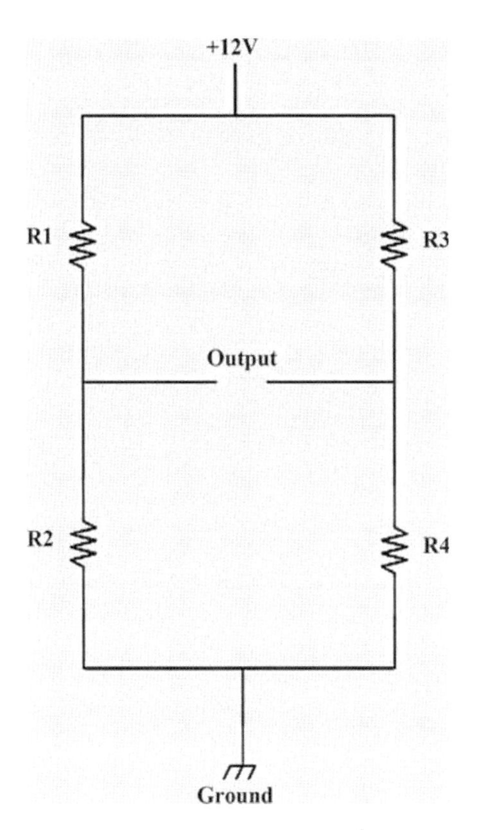

Fig. 8.19. The basic layout of a bridge circuit.

Consider now the current flowing through the output. The voltage at the center of R1 and R2 is 6 V, and the voltage at the center of R3 and R4 is 6 V. Without a difference in potential here, no current will flow. If we place a current meter here, we can detect this null condition when the meter reads zero.

Now consider replacing R4 with an unknown resistor that we think is about 75 Ω, but let's say it is actually 95 Ω. R3 is set to be 75 Ω, the value we want R4 to be, and R1 and R2 are equal, say 50 Ω. Now the voltage between R1 and R2 is 6 V, but the voltage between R3 and R4 is given by:

$$V_{out} = V_{supply}\left(\frac{R_4}{R_3 + R_4}\right)$$

which is

$$V_{out} = 12\left(\frac{95}{75+95}\right)$$

which makes

$$V_{out} = 6.7 volts$$

Introduction to RF Electronics

Therefore a small current will now flow through the output from the high voltage side to the low voltage side. Note here the value of the current is not important; any current flowing means the unknown value of R4 does not match the known value of R3. If R4 is adjustable, it can be set to provide a zero current in the output, and we know it now matches R3.

So far we have been discussing the bridge performance with a DC power supply. An antenna operates in AC conditions, with radio frequency signals. Let's suppose we now connect our antenna in place of R4 but still use a pure resistor as R3. True, R4 will be a reactive impedance this time, and R3 is pure resistance, but it does not matter. It will work adequately well for our purposes. However, this time we must replace the DC power supply with a radio frequency signal from an RF signal generator.

In building a practical test instrument, we can't be sure which way the current will flow through the output or meter contacts. So now consider the circuit in Fig. 8.20. The addition of a series of small signal diodes will ensure the current will always flow through the meter the same way. The connections for R3 and R4 can be fed to chassis-mounted RF sockets, for easy access to hook up the antenna and a reference impedance. If the R3 reference was made to be a variable resistor using a 250 Ω potentiometer, then we could balance the bridge and measure the resistance of the potentiometer to find our antenna impedance. Better still fit a calibrated scale to the potentiometer so resistance could be read directly. The meter is better built in, or you may use a multimeter if suitable connection points are built into it. It will need a reading scale of 50–100 μA. All resistors should be of composition or metal film type and not wire-wound. In practice it is easier to obtain 51 Ω resistor values for R1 and R2.

When constructing the bridge, mount it inside a metal box, with the ground points connected to the case and to the bodies of the RF connectors. Keep all component leads as short as possible. It will certainly work well enough up to VHF frequencies.

The photograph in Fig. 8.21 shows a variable "standard" resistor mounted in a small hand held metal box and connected to the main bridge via a coaxial cable. This version used a 50 μA meter. The lower connector is for inserting an RF signal, the upper two for connecting the standard impedance and the unknown impedance. The pictures were taken before the variable standard was calibrated, and labels fitted. The control knob was taken from a surplus faulty power meter. From start to finish, including cutting the holes in the box, it took a morning's work to build. Not only can it be used to measure antenna impedance, but it can also be used the same way to measure unknown inductor or capacitor values, if a series of known standard inductors and capacitors are available. All in all a low cost but very useful tool for the workshop!

When you come to calibrate the variable standard impedance, use the bridge to do it, and fit known values to the other connector port. Try to have a precision resistance decade box to set it up.

The picture of the circuit is a little hard to follow, but it was constructed using the dead bug style. Note that there are five island pads soldered to the base. The islands are double-sided copper board; the lower side is first tinned with solder and then held onto the base with a small screwdriver while a "wave" of solder is run down the sides, which creeps under and sticks the island to the base. The baseboard is the ground plane. The board is held down by a mounting stud from one corner of the meter and ensures the case is also earthed. The bodies of the RF connectors also reinforce the ground connections to the box. Small coaxial cables

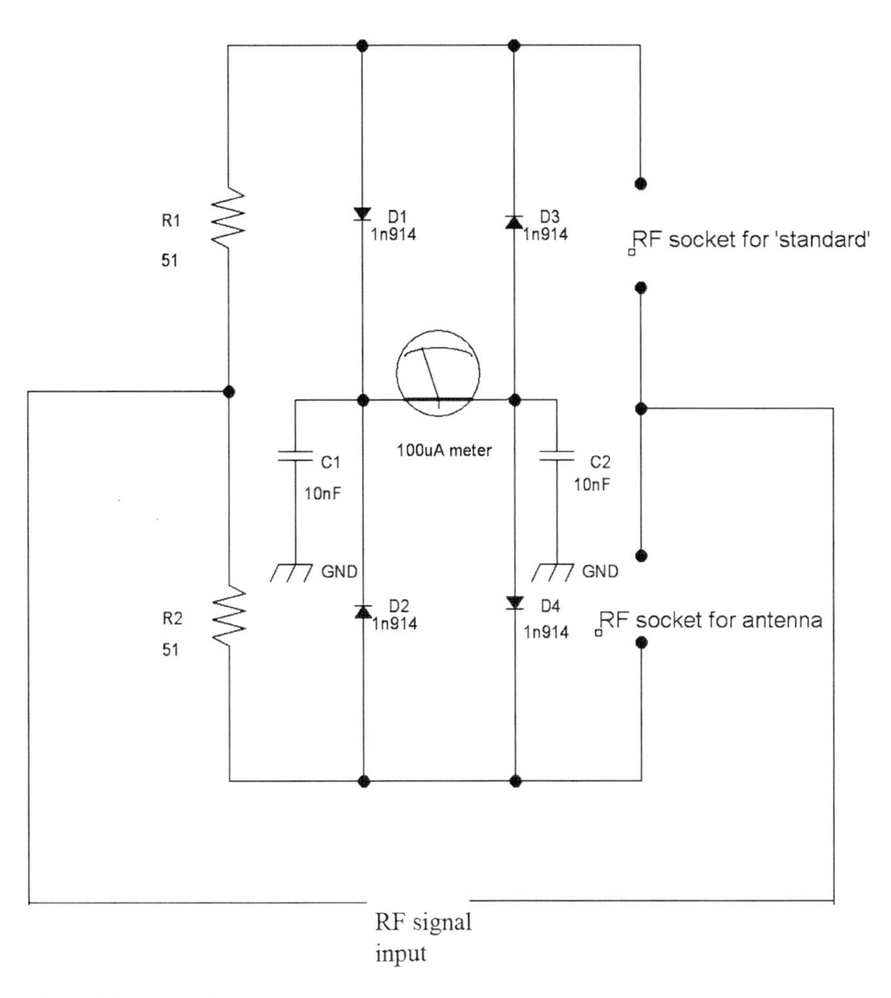

Fig. 8.20. Impedance bridge.

connect the circuit to the meter and the RF connectors. Don't worry about how your circuit looks, so long as you follow the simple guidelines and connect everything the right way.

Constructing the Noise Source

Calibrated noise sources are expensive bits of test equipment. Even surplus equipment units can fetch a high price. Untested surplus units may not even work, or may need repair and recalibration which further bumps up the price.

Following is an interesting way of constructing a noise source, based on articles written by G. W. Swenson years ago in *Sky & Telescope* magazine. It uses a light bulb as a source. The filament of a light bulb is essentialy a resistor, and a hot resistor makes a good approximation to the sort of noise received from space. The fact that

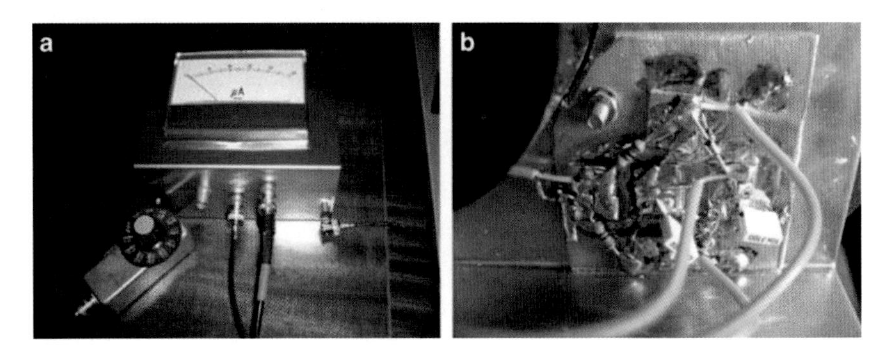

Fig. 8.21. (**a**, **b**) A basic but practical impedance bridge.

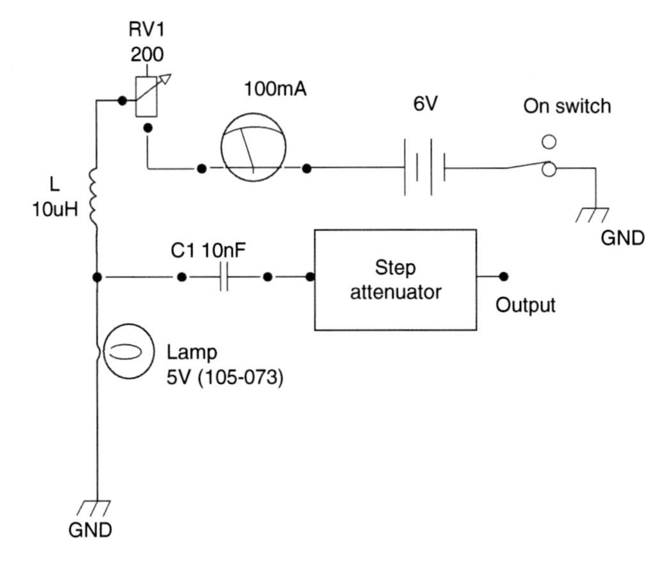

Fig. 8.22. Noise source.

the light bulb filament is hot by virtue of its function saves us the trouble of heating a normal resistor and maintaining its temperature.

In theory, by varying the current through the lamp, its noise temperature will also vary. However, it turns out the response is non-linear, and different between different bulbs. What G. W. Swenson found out was that at rated current, all tested lamps had a noise temperature of 1,300 K. So if the bulb was run at constant current, and an attenuator used to vary the output, we could provide a variable noise source from low temperature up to 1,300 K. Swenson used a 3 V pilot lamp with a rated current of 60 mA. However it is hard to find one of these now, so in chosing a lamp for the job, find one that has about a 50 Ω resistance. This can be worked out from Ohm's law, in that if the rated current is 90 mA and the operating voltage is 5 V then the filament resistance will be around 55 Ω – close enough. At the time of writing the RS components part number 105-073 seemed suitable.

The circuit for the noise source is simple and shown in Fig. 8.22. The lamp should be soldered directly into the circuit and not fitted to a holder, to avoid extra attenuation.

Table 8.6. Calibration table for the noise source.

Attenuation (dB)	Noise output (K)
0	1,300
3	800
10	390
30	290

By adjusting the 200 Ω potentiometer RV1 until the meter reads 90 mA the lamp will be operating at its rated current. If you chose another lamp, the current may be slightly different but should be the rated current. The actual value required for the resistance of RV1 is about 83 Ω, so in fact a 100 Ω pot may be used. However, the low value potentiometers can be a little hard to find, so you can scale a higher value pot down by adding a parallel resistor, so a 500 Ω potentiometer in parallel with a 330 Ω resistor will give you about a 200 Ω pot. As for power, either a 6 V battery or a 5 V external power supply could be used. Once again build it in a metal box, where the ground points are connected to the case, the component wires are kept very short, and the output fed to a coaxial socket whose body is also grounded to the case.

The output of the noise source is varied by using a step attenuator of the type described next. Varying the current is not a reliable way of adjusting the noise level. Remember the lamp will show a 1,300 K noise level at its rated current into a 50 Ω load. Table 8.6 gives the values expected by adding attenuation factors to the output.

It should be noted the lamp characteristics will likely change with usage. Atlhough it is not going to be used for extended hours, the lamp should be periodically replaced after say 100 h of use.

The Step Attenuator

I was lucky to find a good used step attenuator on eBay a while back for £40. I remember thinking it was expensive at the time, but it probably wasn't that bad. Recently I have seen similar ones for £100, and one item brand-new that would operate at up to 40 GHz was £2,400 – ouch!

My own example is good up to 1 GHz, and that is enough for me anyway. However, you can build your own from a few DPDT switches and a set of precision 1% resistors.

The resistors are laid out as a pie network, as illustrated in Fig. 8.23. Each step of the attenuator contains a unit like that; the resistors required for several common attenuation values is given in Table 8.7.

The circuit showing the first three sections of the attenuator with switch contacts is shown in Fig. 8.24.

The switches are DPDT (Double Pole Double Throw) toggle switches. Each switch has two independent contacts or poles, and the double throw is acting as a changeover switch. The RF lines are connected to the common center pins. One side of the switch is shorted with a wire between the pins, and the pie networks are connected to the other side of the switch. In this way each attenuator can be switched in or out without breaking the continuity of the chain. The value of total attenuation obtained is the sum of the individual attenuators. Mount a chassis RF socket at each end for easy connection to equipment.

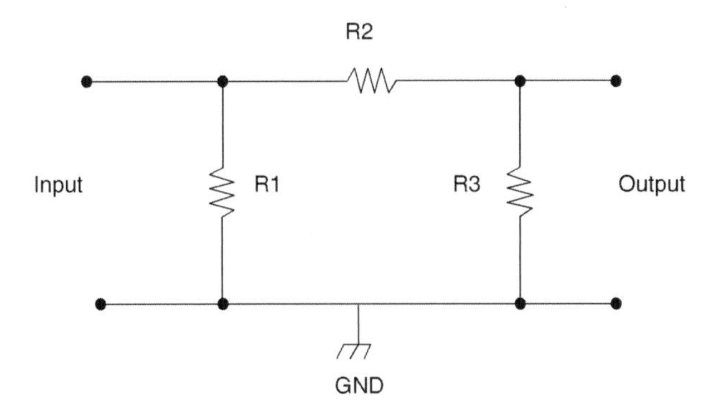

Fig. 8.23. The general form of each stage of a step attenuator. Refer to Table 8.7 for the resistor values.

Table 8.7. Resistor values in ohms for different attenuation steps

Attenuation (dB)	R1	R2	R3
1	866	5.6	866
2	432	11.5	432
3	294	17.4	294
5	178	30.1	178
10	95.3	71.5	95.3
20	60.4	249	60.4
30	53.6	787	53.6

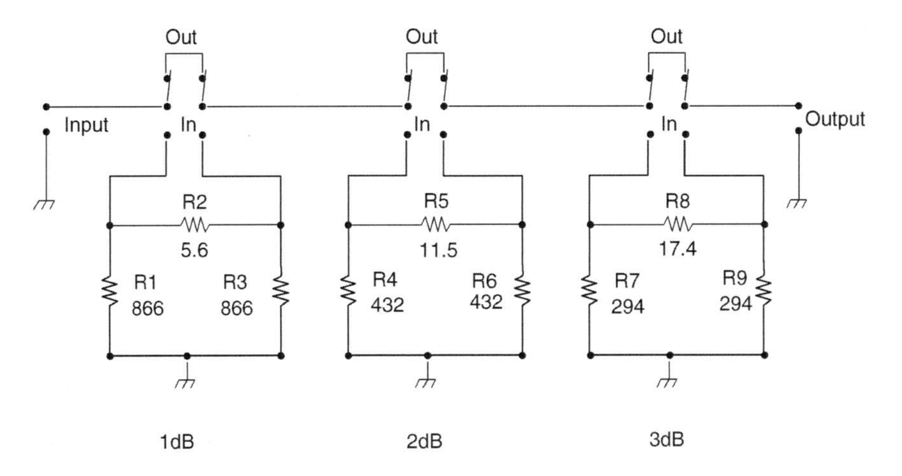

Fig. 8.24. The step attenuator circuit diagram.

Note that a grounded screen should be installed between each attenuator. Pieces of plain copper clad circuit board could be used and soldered together to form an array of boxes within a box.

The values of resistors shown are available in precision ranges; some will cost more than others, depending on their precision and rarity. All leads must be kept short. A resistor will work adequately well up to UHF frequencies of around

450 MHz or maybe more with reduced accuracy. At 450 MHz it would be expected to be within 2 or 3 dB of the maximum attenuation value of 71 dB. At zero attenuation, it may have a stray value of something like 0.4 dB.

Using Your Newly Built Test Equipment

Measuring Antenna Impedance

In order to use the impedance bridge, a signal generator is required that will operate at the frequency for which the antenna was designed. If that is below 200 MHz then it would be fairly straightforward to build an oscillator along the lines of circuits outlined elswhere in this book. However, constructing oscillators for higher frequencies becomes quite difficult, and some experience will be required. If you browse the web pages of Minicircuits at http://www.minicircuits.com/ you will see readymade oscillators covering a wide range of useful frequencies. It is far easier and not too expensive to find a suitable unit for any frequency under 1 GHz (construction at higher microwave frequencies does require some considerable skill). Their range of VCOs is very useful. VCO stands for voltage-controlled oscillator. All VCOs require is a low-voltage fixed power supply and a low variable DC voltage to control the output frequency.

An alternative to this would be to buy a good RF signal generator. You might want to have several, covering frequencies up to 250, 500, and 2 GHz, all generally available from various surplus sources relatively cheaply. They are always useful to have in the workshop. However you chose to generate your RF signal you will probably have to turn up the output power to its highest level (this will not be much, as many tests use milliwatts of power or less).

Note that when using the impedance bridge to test an antenna using a balanced feed (twin line), a balun such as the coaxial balun will be needed to attach it to the bridge. You will therefore need to account for the 4:1 impedance ratio of the balun. The actual impedance of the balanced antenna will be 4 times that obtained from the bridge.

When you connect the antenna to the bridge, say via the coaxial balun, and if you expect the impedance to be 75 Ω (which is confirmed by the bridge with a good deep null reading of close to or equal to zero at 75 Ω), then all is probably well and the antenna is an excellent match. In fact if the impedance is within even 20% of the expected value this would be considered a good match.

If the reading is more than 20% off target, first check that the balun is good by removing it and soldering a 300 Ω resistor across the antenna terminals, testing the balun on its own to confirm that it is indeed 75 Ω. If not then there will be either bad solder joints or the length of the coaxial cable is wrong for the frequency used. Don't forget to account for the velocity factor of the line in determining its length. If all is well with the balun, then work systematically up the line, moving the balun up each section as you go, testing each section.

Once everything has been checked, if no faults were found (such as bad joints), and the impedance is still off, then it comes time to adjust where necessary until the impedance is correct (such as gamma match adjustments and correcting impedance matching sections to suit).

Using the Noise Source to Test Amplifier Performance

Earlier we discussed the importance of the noise characteristic of the first stage of a radio receiver. It is this first stage that contributes the greatest proportion of the overall noise. The test instruments described here are likely to be only useful from mid-UHF frequencies or lower. The procedure and the crude instrumentation will only give us a rough and ready answer, but firstly it will be very instructive to the beginner, and secondly it should be good enough to highlight major faults, or give some confidence in the proper function of the devices.

Set the noise source to a low setting via the attenuator and connect it to the receiver being tested. Assuming a reading is achieved on the output make a note of the value and the temeparture setting of the source. If no output is obtained, increase the noise level slowly until it reads something like mid-scale, or a decent level anyway. Now add a 3 dB attenuator between the noise source and the receiver (this is an additional attenuator, not a step of the step attenuator in the noise source). Adjust the noise source to achieve the same output level as before. The difference in the two temperature settings is then a measure of the noise performance of the receiver.

An additional check can be done. Take a short piece of coaxial cable of the type used to feed the receiver, solder a resistor whose value is the same as the impedance of the cable across the end of the cable (keeping the leads short), and fit a suitable coaxial connector to the other end. Seal the resistor with silicone sealant and let it dry.

When ready, fit the resulting "probe" to the receiver, and dip the sealed resistor end into a bowl of ice water that has a lot of ice cubes in it and which has been stabilized to 0°C. Take a reading from the output of the reciever. Now dip the resistor into a pan of boiling water that is at 100°C. The output of the receiver should increase noticeably. If it does it should have sufficent sensitivity to do some useful astronomy.

Chapter 9

Building a Very Low Frequency Solar Flare Monitor

This instrument is a SID detector. SID stands for solar ionospheric disturbance. The project is very easy, and remarkably effective, so it is recommended that beginners start here. The success you gain will provide you a boost and encourage you to continue in this fascinating hobby. In fact there is virtually no electronic construction involved! Apart from a suitable aerial, everything is off the shelf and you probably already own it – a PC with sound card!

Firstly let's introduce the science behind the telescope. In the strictest definition, the VLF (very low frequency) receiver is not a radio telescope. It can't detect radio waves directly from space. This is because Earth's Ionosphere is opaque to radio wavelengths longer than about 20 m most of the time. Propagation of manmade VLF radio signals over long distances depends on the electron density and particle collision rates in the lowest of the ionospheric regions, the D layer.

In daylight ultraviolet exposure causes ionization of nitrous oxide molecules in the D layer. However, there is a high rate of recombination, where the free electrons join once again with the molecules. The net effect is a relatively low level of ionization but a quite stable level during the day. At night, with the absence of solar UV, and X-ray the degree of ionization drops even lower, although it does not completely disappear. Galactic cosmic rays help to generate some free electrons, energetic particle collisions from the E layer above account for the rest. Therefore at night the state of ionization is much more random and variable, and where the region begins at a higher altitude.

Now for the useful bit. During solar flares, the Sun emits bursts of X-rays that significantly and rather quickly enhance the ionization of the D layer. This results in a measurable change in the strength of received VLF signals. Thus it is possible to use a VLF receiver as a solar flare monitor. Clearly it would be hard to distinguish between changes in output power of a VLF transmitter and a genuine SID. However, the characteristic patterns are different, and experience helps allow the observer to detect the difference. As a safeguard against manmade effects (some transmitters tend to go off periodically), this experiment is capable of monitoring several transmitters at the same time.

J. Lashley, *The Radio Sky and How to Observe It*, Astronomers' Observing Guides,
DOI 10.1007/978-1-4419-0883-4_9,

Manmade VLF signals, for our purposes below 100 kHz, reflect from the D layer and back off the ground, making a series of hops over large distances. The D layer quite strongly absorbs higher frequencies until it once again becomes transparent in the VHF. The question is, what happens to the received signal when a SID occurs? This is more complex than you may think for short- to middle-range observations. If at your location you receive the signal directly (the ground wave) as well as receiving the signal by reflection (the sky wave), due to path length differences the two interfere to generate the result. This interference could be constructive, leading to enhanced strength, or destructive leading to weaker strength. So, during a solar flare the enhanced reflective properties of the ionosphere could result in slight change in the altitude at which the signal is reflected, also affecting the path length of the sky wave. The resultant interference pattern could then increase or decrease the received signal strength. It would be an interesting exercise to monitor as many VLF channels as you can to study the effects for yourself. For longer range stations, the ground wave will be weak or not present at all, and since it is only the sky wave that is affected by solar activity, these channels will often provide the greatest responses.

Construction of the Antenna

For this experiment, the antenna is the only item that requires construction. It is very easy to build, and so this device will provide fast results. You can have a working solar flare monitor in a single day.

The most suitable antenna at very low frequencies is a multi-turn loop. It does not matter whether the loop is round or square, in fact it is much easier to make it square. Fine-grained hardwoods offer better strength, but sections of knot free soft wood would also work.

Cut a piece of wood 57 cm long from a piece of 15 mm square wood. (The size is not critical, but keep it small and close to this size.) This will form the upright of a cross. Cut another piece to a length of 52 cm to form the horizontal section of the cross. The vertical section is longer to allow it to be mounted onto a baseboard.

Next cut four pieces of the same material to a length of 60 mm. Mark a line across each of them 1 cm from one end, and make a saw cut to a depth of half the thickness, completing the cut down from the end to remove the piece, leaving a tongue 1 cm long and half the thickness of the wood on all four short pieces.

Next mark the center of the 52-cm-long cross member, then make two more marks across the width of the material 7.5 mm (adjust accordingly to half the thickness, if you have chosen a different size stock) either side of the center line. Put two saw cuts down carefully to half the thickness of the wood, taking care to cut on the inside (towards the center) of your pencil lines. Using a small sharp wood chisel cut away the material between the saw cuts leaving a flat bottomed notch. Do the same on the longer upright, but don't forget it's not now in the center but offset 50 mm towards one end.

By now fitting the two long sections together in the center and gluing the joint you have a cross framework. The small pieces are now glued to the tips of the cross with the cutaway section on the inside forming a groove to carry the copper wire loop. The bottom piece needs to be carefully placed by measuring 52 cm from the top to the outside, not to the base of the notches. Once the glue is dry, the cross should have a 50-mm extension at the base, which can then be glued

Fig. 9.1. The VLF loop antenna. It is a square with 400-mm-long sides and made up from 125 turns of enameled copper wire. There is a tuning capacitor on this one, which is not needed for this project.

into an appropriate hole in the center of a piece of 12- or 18-mm plywood approximately 30 cm^2.

The last stage in construction involves winding 125 turns of copper wire on the frame. Use enameled copper wire, of 24 or 26 swg and approximately 0.5 mm diameter, purchased from an electronics supplier. Try to get enough on a single roll to complete the job so you won't have to join it. You will need about 185 m. If you have to join two sections, be sure to remove a small amount of the insulating enamel from the ends to allow it to be soldered. You could paint a little enamel or varnish on the joint afterwards to ensure the turns remain insulated from each other.

The two ends of the loop can now be soldered to a Phono or a 3.5-mm jack socket mounted into the baseboard. For the purposes of this experiment, the loop works very well untuned, so no capacitors are needed. Although in the prototype antenna shown there were fitted some fixed capacitors and a variable air spaced capacitor to allow the aerial to be tuned to the frequency of the channel, the main reason for them was to connect to a borrowed hardware receiver. In practice there is no difference in this experiment if there were no capacitors. Although a dedicated tuned receiver can give excellent results, it means you need several receivers, and a multichannel data logger to monitor more than one channel at a time. Read on and find out an easier way to monitor multichannels.

The Computer as a Receiver

Now for the clever part – using an appropriate sound cable, which connects the antenna to the microphone port on your computer sound card. Download and install a piece of software called *Spectrum Lab* from the following website: http://freenet-homepage.de/dl4yhf/spectra1.html.

The software was written by Wolfgang Buescher, whose amateur radio call sign is DL4YHF. It is free to use and is an extremely useful piece of software for many radio projects.

Setting Up Spectrum Lab

When you first run *Spectrum Lab*, the window will appear as in Fig. 9.2. First go to the Start/Stop menu and click on "Start Sound Thread." The items "Spectrum Analyser #1" and "Audio Input: from ADC, active" should already be ticked.

With the antenna connected to the microphone port of your PC, and with the sound card recording control set to microphone, you should see a plot that looks something like Fig. 9.2.

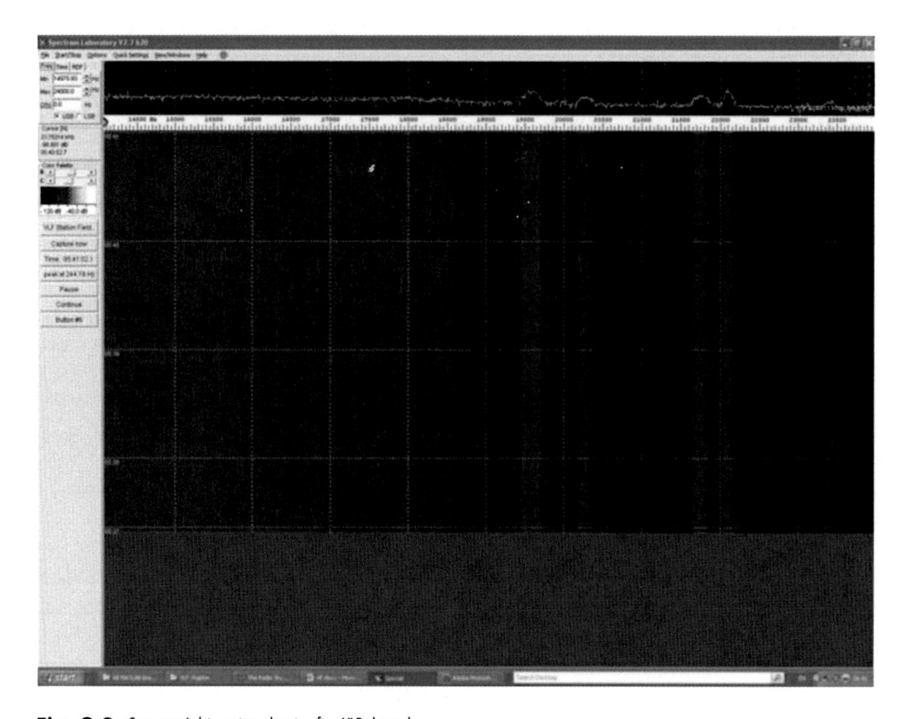

Fig. 9.2. *Spectrum Lab* in action, showing five VLF channels.

The upper graph is the audio spectrum. The lower graph is referred to as the waterfall. The broad vertical lines on the waterfall and the corresponding peaks in the spectrum plot should be VLF radio stations. However, some care needs to be taken before you jump to that conclusion. Some interference looks exactly the same as a VLF channel, especially that of the old CRT tube televisions at a frequency of 15,625 Hz. The plot in Fig. 9.2 does not show this; it was taken at 6:45 a.m., when few people watch TV!

Ok, now we want to tweak the display to suit our needs. A larger spectrum display is best. Go to the options menu and click on the Spectrum display settings item, which will bring up a window as in Fig. 9.3.

On the right hand edge half way down is an item Spectrum graph area (pix), which refers to the height in pixels of the spectrum plot, next on the left side of the main window you will find a color palette with two sliders for B, brightness, and C, contrast. By adjusting these you can enhance the visual appearance of the waterfall. This is merely a preference feature that has no effect on the values of the signal strength.

The most important feature of *Spectrum Lab* for our purposes is setting it up to log data of the signal strength to a file for long-term monitoring. There is a neat built-in plotter for that purpose. Before you configure the plot window you need to set the sampling rate for the sound card to as high a value as your card will support. Most built-in sound cards will have a maximum sampling rate around 48,000; some may be 96,000. Whatever sampling rate you choose, the Spectrum plot window will show frequencies up to half that value. So for a 48 K/bit sampling rate the plot window can show up to 24 kHz radio channels. Set the sampling rate by going to the Options/Audio settings menu item, and under the audio processing section there is a drop down box for setting it.

To open the plotter function go to the View/Windows menu item and select the watch list and plot window lower down in the list. Next open the file menu on the main window, and select "load settings"; it should then list a variety of configuration files in the configurations subfolder, one of which is "VLF_Station_Plotter.usr." This will bring up a box similar to Fig. 9.3.

This shows a list of VLF stations currently found in Europe; for other regions the plot can be modified easily to suit.

The first tab is the watch list, which defines the frequency and the range over which it will display. By double clicking on each item it can be edited. The Title is simply a useful text string. The expression column defines what is going to be measured. Here peak refers to the peak amplitude, and in the parentheses is the range of frequency it will monitor – low first, high second. When changing the frequency to suit your application, use the cursor to determine the start and the end frequency to monitor in the main window. As you point the cursor to the spectrum the frequency value appears beside the cursor. These values can then be typed into the expression. The software will automatically determine the peak value between these two frequencies. Be warned if no transmission is active in the range; it will still record something, the peak noise!

The result column is generated by the software; the format column should not need to be changed. The scale maximum and minimum refer to the decibel range of the vertical axis. You may ultimately want to lower the max level, depending on the peak strength of the signals you receive. You might want to lower it by 50 dB, for example. By properly scaling the plots to fit your screen window you can maximize the ability to spot changes in signal strength that could be due to solar flare activity.

a

SpecLab Configuration and Display Control

TRX Control | Memory | Filenames | Wave Files | Markers | System | Freq-Resp
Spectrum (1) | ..(2) | ..(3) | ..(4) | Radio DF | FFT | Audio I/O | AD/DA Server

Vertical Frequency Axis
double-width waterfall lines
optimum waterfall average
triggered Spectrum more..
peak detecting cursor
emphasize MIN+MAX values
show spectrum as bargraph
Amplitude Grid (dB or %)
one pixel per FFT bin
multi strip WF, 100 pix/strip
non scrolling WF
peak holding graph, hold time (s): 5
long-term average clr
Show: both / Plot right
+ amplitude bar
Maths: none
Spectrum graph area (pix) 600
Channels / Connections ...

Waterfall Scroll Interval
300 ms sec minutes
automatic (50% overlap)
smooth scroll, high CPU load

Waterfall Time Grid
enabled
Interval 60 sec min
Source (expr)
Style dotted lines
user-defined time label format :
Labels hour:minute
YYYY-MM-DD hh:mm:ss

More spectrum display settings on the next >> and on the "Radio Direction Finder" tab >>>

Shown: Settings for Analyser 1, channel 1 (L)

Apply | Close | Help

b

Spectrum Lab Watch List and Plotter

File Settings Plotter View Help

Watch List | Plotter | Layout | Horizontal | Channels & Colors | Memory, Misc. | Export

Nr	Title	Expression	Result (Value)	Format	Scale Min	Scale Max
1	Noise	noise_n(8000,20000)	-90.4	##0.0	-130	-20
2	TVsync	peak_a(15620,15630)	-77.8	##0.0	-130	-20
3	16k4	peak_a(16350,16450)	-79.0	##0.0	-130	-20
4	18k1	peak_a(18050,18150)	-74.9	##0.0	-130	-20
5	18k3	peak_a(18200,18400)	-79.5	##0.0	-130	-20
6	19k6	peak_a(19500,19670)	-62.3	##0.0	-130	-20
7	20k27	peak_a(20170,20370)	-75.8	##0.0	-130	-20
8	20k9	peak_a(20810,20990)	-91.2	##0.0	-130	-20
9	23k4	peak_a(23350,23450)	-84.5	##0.0	-130	-20
10	15k98_qr	peak_a(15900,16100)	-75.4	##0.0	-130	-20

(cell info)

Fig. 9.3. (**a**) Spectrum options. (**b**) Spectrum watch list.

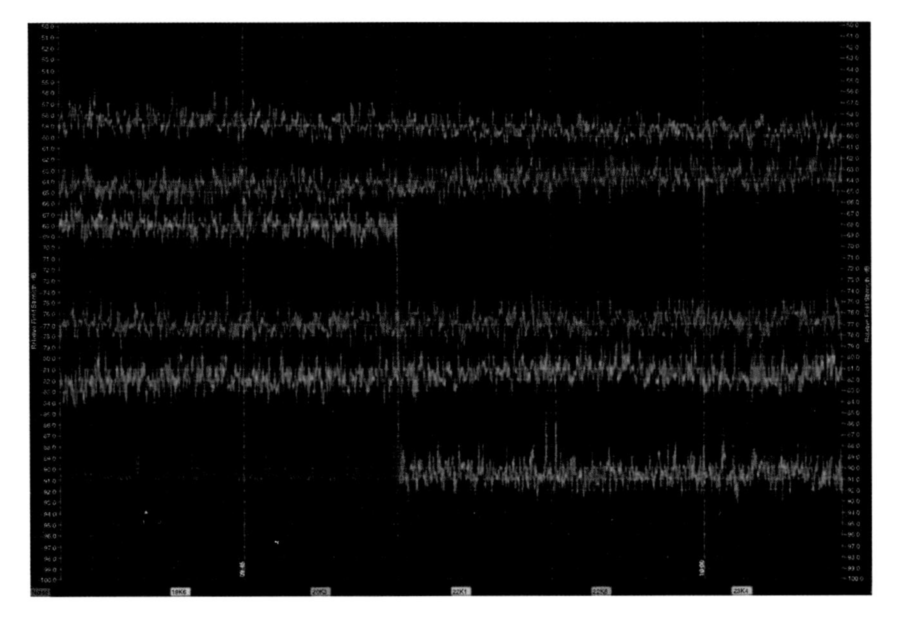

Fig. 9.4. VLF plot window from *Spectrum Lab*. Note how the 22.1 kHz (*green*) channel was turned off at 09:50, and the signal dropped to the noise floor (*red trace*) at −91 dB.

Once you are happy with the definition of the channels you want to watch, go to the "Channels & Colors" tab and turn off any channels not required. To do so select the channel number, and go to "graph styles" and select off.

Finally go to the export tab, and set a filename for the log followed by selecting the check box "periodically export the plotted data." Finally select the radio button "comma as column separator."

Go to the file menu of the watch list window and select save settings, and save it. To operate the plotter, from the plotter menu, click on "Run (now stopped)." It will change to "Stop (now running)" with a tick beside it. When you select the plot window you will see a slow-moving graph start to develop. You should now be observing and recording your observations! Daily data can be exported as a text file for later processing if needed.

Figure 9.4 shows a plot screen obtained from *Spectrum Lab*. It covers about 20 min of time and was a series of 1-s samples. In practice samples should be every 15–30 s. Note how noisy the trace is on a short time scale. Should a particular plot contain useful information, such as a solar event, the data can be smoothed by later processing.

Figure 9.5 shows raw data exported via a text file to Microsoft Excel and plotted. This is data that was gathered on July 17, 2009. The raw data is very noisy and confused. Figure 9.6 shows the result of averaging samples over a 5-min period.

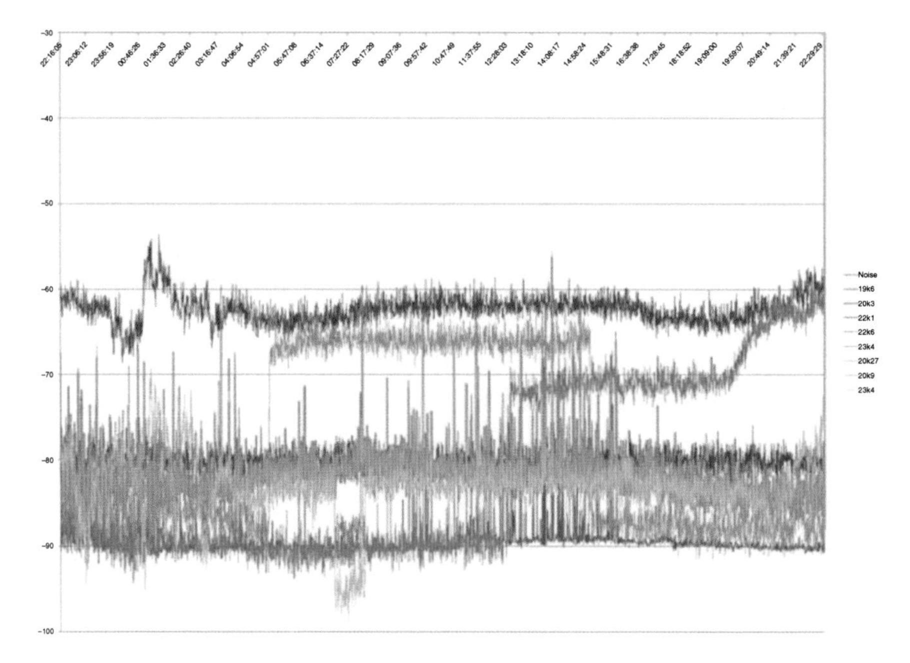

Fig. 9.5. VLF plot of five stations plus the noise background for July 17, 2009. This is the raw data showing lots of short period noise in the traces.

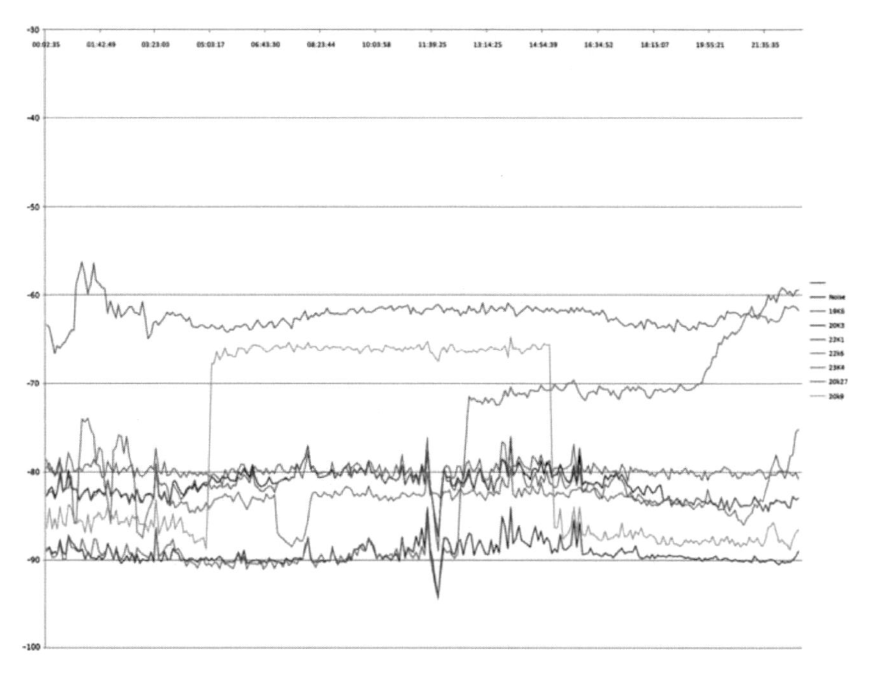

Fig. 9.6. The same data as Fig. 9.5 but averaged over 5-min intervals. Note the two channels that switch on and off during this period. There is a significant dip in one trace, but this is not seen on the others, so it is not likely to be a solar event.

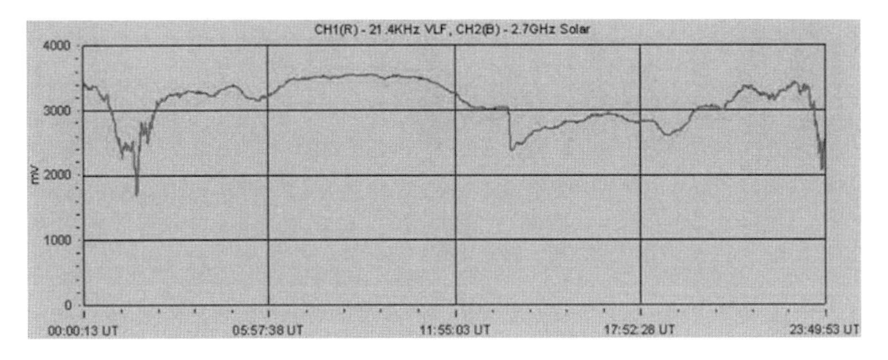

Fig. 9.7. VLF channel plot dated July 9, 2007, showing a significant SID event at around 13:00 UT. Note only channel 1 (VLF) is shown in this graph. (Image courtesy of Martyn Kinder.) Note also the wide variations in signal at night. There are always characteristic dips in the signal at both sunrise and sunset daily.

Figure 9.7 shows a VLF plot taken with a hardware receiver now available ready built or as a Kit from UKRAA, an offshoot of the British Astronomical Association Radio Astronomy Group. This was taken on July 9, 2007, and shows a SID event at about 13:00 UT. The data was captured using a homemade data logging interface based on the Maxim MAX186 chip (see Chap. 13) and *Radio Skypipe* software.

Chapter 10

Microwave Radio Telescope Projects

The first project is a version of the Itty Bitty radio telescope designed originally by Chuck Forster of SARA (Society for Amateur Radio Astronomy) and outlined on the teacher's resource pages of the NRAO website.

This project will take the instrument to the next level. For example, we will add a computer interface for data logging. The heart of the telescope is a satellite finder and a universal satellite television LNB. The block diagram of the telescope is shown in Fig. 10.1.

The aims of the project are the following:

- To learn the basics of microwave radio receivers.
- To gain experience building real systems.
- To gain experience using a radio telescope.

Required Parts

The Dish

The dish can be any surplus or even a new satellite television dish. Even the smallest oval dish is capable of solar observation. Although a larger dish will improve signal strength, since the aims of this project is primarily educational, it is not recommended to use too large a dish at first. A 45 cm to 1 m size is best and is easy to handle and mount. If the dish is kept small the system can be made portable and used to demonstrate radio astronomy to friends, clubs, and at star parties.

Dishes come in two main types, offset and concentric. The offset type is very common in the small sizes, but poses more of a problem in accurately aligning the dish for practical observing sessions. The idea of an offset dish is to make it easier to fit to a house wall for use in a television receiving system. The dish sits nearly vertical against the wall but is receiving its signal from a satellite at an angle at about 20° to the horizontal. The LNB is also offset at an angle on the opposite side, so that the reflected signal is received. In radio astronomy we may want to point the dish to any arbitrary position, but it's always a bit tricky to know where it is

J. Lashley, *The Radio Sky and How to Observe It*, Astronomers' Observing Guides,
DOI 10.1007/978-1-4419-0883-4_10,

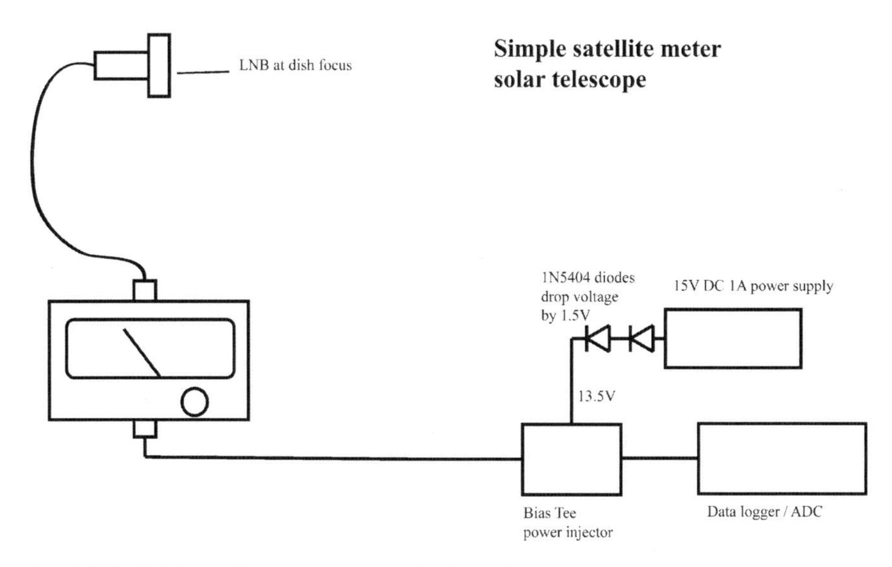

Fig. 10.1. Block diagram of the basic microwave telescope. The dish is not shown.

pointing. It's really worthwhile looking for a small concentric dish. Typically small concentric dishes are not used for television receiving but for microwave communication links or weather satellite receiving or such. These will probably need some conversion in order to mount a satellite television LNB, but more on this later.

The LNB

LNB stands for low noise block. The sky background is fairly quiet in the lower to middle microwave bands. It is therefore possible to utilize low-power satellite signals for direct reception of television in homes. Due to the weak signal strength at Earth's surface, a low noise amplifier is required to properly detect it. Remember that the most critical portion of a radio receiver that influences the overall noise performance the most is the first stage. LNB's are available with a noise figure as low as 0.3 dB. Many of the low cost units offered to the consumer market seem to be optimistic in the specifications. Although 0.3 dB may be the best a given LNB can provide, this may only be true at one small part of its receiving range. The average performance across its bandwidth may be closer to 0.6 dB. Many professional quality LNB's quote noise figures of 0.6–0.8 dB, probably a more realistic figure. Any of these would work well in this project. Note also, LNB's come in offset and concentric types and they should not be mixed.

To understand the workings of an LNB you need to understand some of the terms used in their specification. First, the term universal LNB may not be what you would expect. Universal does not refer to the way it is mounted, such as a universal bracket or clamp. It refers to the frequency coverage, the way that frequency tuning is achieved and in the way its polarity is controlled. Also, universal does not mean it is the same worldwide, either! The notes here refer to the European meaning of Universal LNB.

The frequency bandwidth – the coverage of the Ku band satellite TV allocation – is from 10.7 to 12.75 GHz. A universal LNB can cover this whole range but not

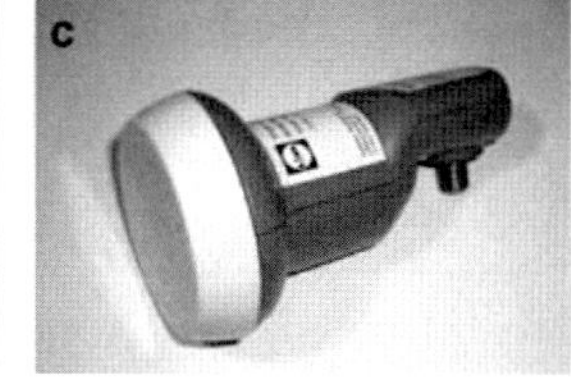

Fig. 10.2. (**a**) The LNB circuit. This kind of circuit is difficult to make at home. Note that the cover has built-in shielding walls to isolate various parts of the circuit. (**b**) LNB with a c120 feed horn used on concentric dishes. (**c**) A European Universal LNB used on offset dishes.

simultaneously. The bandwidth is broken down into two parts, 10.7–11.7 GHz and 11.7–12.75 GHz, sometimes referred to as low band and high band. Traditionally only low band was used, but to add extra space for more channels the TV allocation range was extended. Universal LNBs employ a 22 kHz tone signal to switch it to the high band. This is normally provided by a satellite TV tuner box. In this experiment the tuner is not present, and without a 22 kHz tone generator the LNB will default into operation in low band, providing our receiver with a 1 GHz bandwidth (Fig. 10.2).

The Universal LNB has two antenna probes mounted orthogonally, allowing it to switch between horizontal and vertically polarized radio channels. The way this is done is by switching its supply voltage between two states, 13 and 18 V. Actually the LNB will happily operate at any DC voltage between 12.5 and 18.5 V, although there is a gray area between 14.5 and 15.5 where you can't be quite sure what polarization it is using. The lower voltage makes it work in vertical mode, and the higher voltage in horizontal mode. Clearly horizontal and vertical refer to its primary role as a TV dish mounted vertically on its stand. In a radio astronomy function, the dish may well tilt at other angles if objects are tracked across the sky. In its simplest form the actual polarity used does not matter, as the Sun emits a random mix of polarities anyway.

Now, looking inside an LNB, we can see that it provides two functions for us. First it significantly amplifies the radio signals detected by the probe. Second it translates the frequency down from the Ku band to the L band, which is the first intermediate frequency of the system. This IF frequency is in the range 950–2,150 MHz.

Amplifying and reducing the frequency as close to the antenna as possible avoids most of the problems of loss of signal that would otherwise be present when sending a signal down a feeder to the receiver some distance away. There is a down side to this. The LNB is mounted via a short feed horn right at the focus of the dish. Normally this is no problem, but if we are to observe the Sun for any length of time, the reflected energy, particularly infrared and light, will cause significant heating of the electronics. This is not only detrimental to noise performance but could melt the plastic components of the LNB body and damage the electronics inside! It would be advisable to at least paint the dish with a matte black all-weather paint to reduce its reflective properties in the IR and visible wavebands. A better way to protect the electronics is to build an extended waveguide that curls around the back of the dish out of the way and mount the LNB there. This technique, however, will be left to the reader to think about.

A useful feature of the LNB is that it is powered by the same cable used to feed the radio signal, therefore saving the effort of routing extra power cables. This technique can of course be used for any mast head electronics. Normally the television receiver would power the LNB, but in our case we will have to construct what is known as a bias tee, a device for injecting power into the RF feeder.

In place of a conventional television receiver, we will use a modified satellite finder as our receiver. The basic satellite finder is a readymade signal strength meter. It could be used unmodified, but would then be a visual-only device. By modifying it we can connect a data logger to it to record signal changes. The beauty of the signal strength meter is that we have another layer of amplification, this time at the IF frequency, and a built-in detector and integrator. The output of the integrator is DC, which is normally used to feed the analog meter of the display.

Construction

The dish and LNB are best purchased together. Often the small dishes come as a package, such as the Sky mini dish in the UK and the Direct TV dish in the United States. As mentioned earlier LNB's come in concentric and offset varieties. It is important to match the LNB with the style and focal ratio of the dish being used. This ensures that the feed horn can make best use of the entire dish aperture. An offset LNB will have an oval beam pattern; if this is used in the center of a concentric dish it will prove inefficient, and the edges of the dish will be wasted.

If you are to use a concentric dish that needs modification to fit a TV LNB, it is first important to work out where the focus will be. Clearly this will be in the center, but the critical dimension is how far from the center of the dish is it. This boils down to working out what its focal ratio is. There is a simple formula you can use to calculate the focal length of a concentric dish:

$$f = \frac{D^2}{16d}$$

where f is the focal length, D is the diameter of the dish, and d is its depth. All these are measured in the same units.

D can be found by simply measuring the diameter with a tape measure, but depth needs a long-enough straight edge to cross the face of the dish. The depth is measured from the edge of the dish to the center point. Lay a good straight edge such as a length of metal box section across the center line, and measure down with a tape measure or use a thin rod and mark off the height with a felt pen so it can be measured. Plug the numbers into the formula and work out the focal length. The LNB is placed such that the focal length is a small distance inside the feed horn.

The face of the feed horn is sometimes invisible, as it is covered with a plastic cap. A little educated guesswork may be needed. If you need to know the focal ratio this is of course given by f/D.

The situation is completely different for an offset dish. An offset dish is a portion of a larger paraboloid, in effect a piece cut out of the edge of a larger dish. Therefore you don't know where the center of the dish lies. If you want to know how to work out the dimension of such a dish refer to the excellent article at http://www.qsl.net/n1bwt/chap5.pdf. For this exercise use a matching LNB with an offset dish – no modifications needed.

Next, before we do any changes to the satellite finder, we need to build the bias tee, so the system can be tested at this point. The circuit diagram is shown in Fig. 10.3.

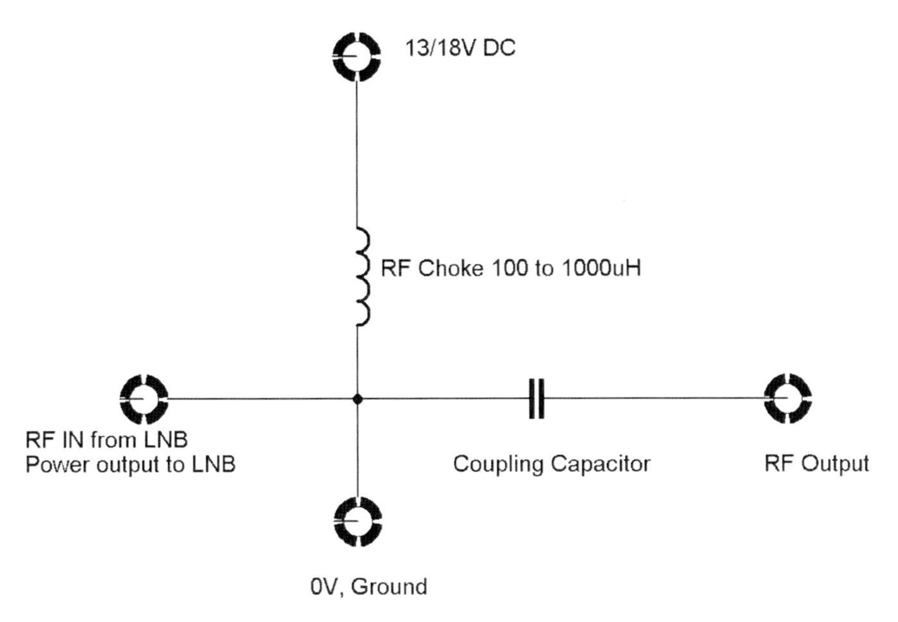

Fig. 10.3. Circuit diagram of a bias tee for TVRO. The ground connection is common to the shield of the RF in, RF out and the 0V from the power supply.

The most important consideration in making a bias tee is that the capacitor reactance should be very low and the inductor reactance very high at the frequency it is used. This allows the signal to pass via the capacitor with little attenuation, while blocking the DC power. The inductor prevents the signal from leaking into the power supply by virtue of its high impedance. The same technique is used inside the LNB and signal meter to separate power from signal again.

Note that the bias tee allows additional equipment to be connected to the RF out port for later expansion and experimentation. It is not strictly necessary to include the capacitor and RF out port for the simplest system.

Figure 10.4 shows the basis of the receiver detector system before any modifications were made. The connector on the right normally goes to a satellite TV receiver, and the connector on the left goes to the LNB.

First test the system in its basic form. Connect the bias tee to the receiver side of the meter, and the LNB to the dish side of the meter. Connect the power and turn on. By increasing the sensitivity control a whistling sound is generated, and the pitch increases with increasing signal strength. Setting the sensitivity somewhere in the lower part of the needle travel and pointing the dish at the Sun causes an increase in signal heard and seen on the meter. With small, careful movements various satellites will also be clearly found along the Clarke belt at zero declination.

Now for the interesting part. Break down the components again so that the meter can be taken apart. In the model of satellite finder illustrated in Fig. 10.4, there was no obvious way of opening the case. But the back panel was simply a press fit, and with a little work popped out. Since the circuit was soldered in situ to the F sockets, you needed cut away the plastic and remove three small screws so it could be withdrawn. Don't count on retaining the same box, however, as it is not shielded from stray RF. Figure 10.5 shows the circuit after removal from the box.

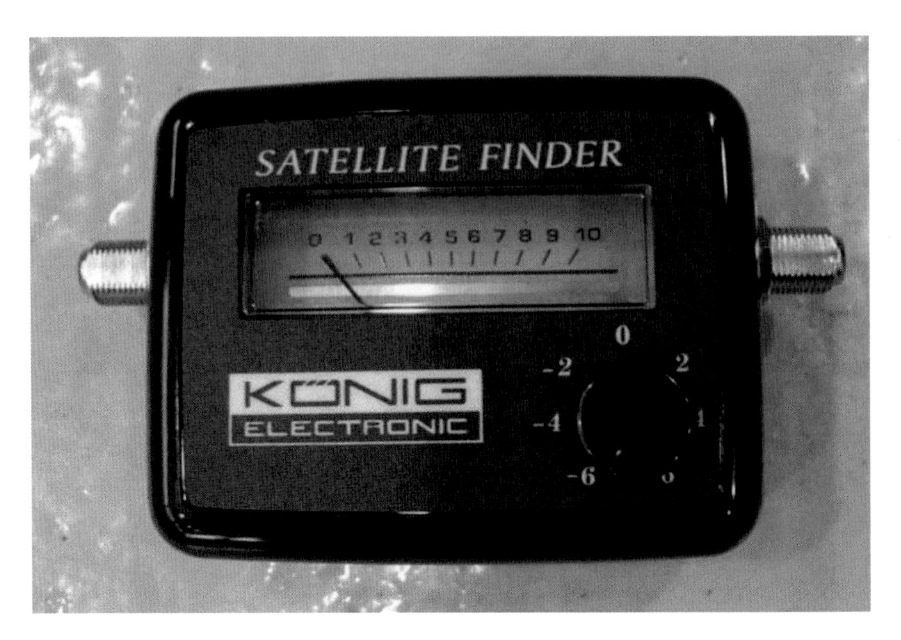

Fig. 10.4. The satellite finder as it came out of the box before any modifications were made.

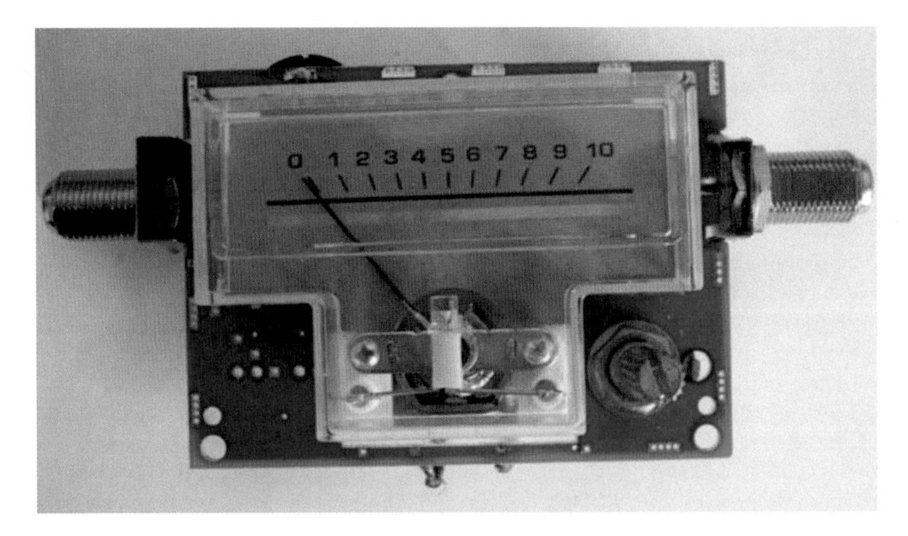

Fig. 10.5. The satellite finder with the box removed.

The first modification involves removing the meter. Simply de-solder the connections on the reverse side and pull it out. Solder a pair of wires to the same points, using the convention of red wire for the positive connection and black wire for the ground wire. Identify the ground wire, which is connected to a large area of copper surrounding all the places where there are no thin traces. If in doubt, connect the power and lnb again and turn up the sensitivity so that a buzz is heard, and use the multimeter to work out the positive and negative terminals.

The next modification is to remove the sensitivity potentiometer. Carefully de-solder it, and use a solder removal tool, such as a vacuum sucker, which can be purchased from electronic component supplier. Get as much of the solder off as you can. Then reheat the terminals and slowly remove the potentiometer. It is very easy to damage the fine copper tracks if you are too abrupt with it. Even trying to be careful, you can still slightly damage a track. To fix it any damage carefully remove a little of the green lacquer insulation, which allows you to solder a fine wire to it. Alternatively follow the trace until the nearest component connection is found, and solder the wire to end of the component. The aim after removal of the potentiometer is to solder in three wires that can be later soldered onto the potentiometer after moving it and mounting it on the outside of a new metal box.

The F connectors need to be removed. Heat the solder points and use the vacuum tool to suck away the excess. Retain them for installation on the new metal case. You can also replace them with N-connectors (if you already have them). Using some short lengths of thin diameter coaxial cable, take off a short length of insulation at the ends and solder into place where the F connectors were. Always keep short lengths of useful cable like this in your "scrap box" from dismantled surplus or dead equipment. Don't worry too much about the cable impedance. The lengths are so short it hardly matters.

Next you can de-solder the buzzer and turn it around 180° and solder it back into the circuit. A fine wire is then soldered onto the board where the second buzzer pin used to be. Another wire, perhaps yellow, can be soldered to the now free-floating buzzer pin, the junction being insulated with heat shrink or electrical tape. The pair will later be soldered to a panel switch so the buzzer can be disabled for long-term observation runs.

A suitable metal enclosure should be purchased, large enough to take the circuit board of the meter and deep enough to allow the board to stand off the metal base and clear the electrical connections in the top. The F connectors (or N's, if you prefer) should be installed into the case lid. The bodies will make an electrical ground connection to the box. Then mount a miniature SPST toggle switch and solder to the buzzer wires. Make a paper template of the circuit board mounting holes, and stick that to the base of the box with tape. Drill the relevant holes to take stand offs. Stand offs can be purchased from electronic component suppliers and consist of a rod which is drilled and tapped with a thread at both ends. This allows the base of the stand off to be screwed to the case, and the circuit board to be screwed to the tops of the stand offs. These ensure the board connections do not come into contact with the box and short out. The mounting holes in a meter can be through the ground plane areas. This means that by using metal stand offs a ground connection is made to the body of the box. If plastic standoffs are used or if you need to insulate the mounts from nearby tracks, then a ground wire will need to be soldered from the board and bolted to the box somewhere to ensure good shielding properties.

A socket of some kind is needed for the output signal (formerly the meter connections). You can use a phono socket. It's not very critical what type is used since the output is a DC voltage.

Now the power feed. While you could use the previous technique of injecting power via an external bias tee, you should think of potential future experiments of placing extra electronics on the output side. So you can remake a bias tee internally, by soldering it directly to the output coaxial socket. A small piece of circuit board can be used to install the capacitor and inductor. This way the capacitor blocks the DC power supply from entering the output coaxial line. This method means that a power socket needs to be installed in the case of an external DC power supply. You can use a pair of banana sockets or a 3.5 mm power-type jack socket.

Once all the connections have been made, recheck everything, and use a multimeter on continuity test or a low resistance range to confirm basic things, like the power connections are not shorted and that the power ground is connected to the metal case. Check that the ground plane (the large area copper pads) of the circuit is also connected to the metal body. Check that the RF connectors are properly grounded to the box, and that the center pins are not.

The only remaining task is to connect the output socket to a suitable data logger. Refer to the last chapter for details on making or buying suitable data loggers and how to use to them. In the example given the maximum output voltage encountered was 7.8 V; you may want to measure yours, as it may influence what data logger you build or buy.

Once everything is connected up, apply power and confirm that the buzzer still operates by turning up the sensitivity. Then connect it to an LNB on a television dish and confirm the system still works as it did before, by making sure the pitch of the buzzer is still changed, and by connecting a multimeter on the DC voltage scale to the phono output socket. If you do not have satellite TV then use the dish to point at the Sun; it does not have to be a clear sky, although you will have to guess where the Sun is in cloudy conditions! Operationally it should be noted there is a jump in output voltage when the buzzer is disconnected. This should not pose any really problems. The absolute voltage reading is not important. It is only the changes in voltage with received signal that is important. Once set up the buzzer plays no further role in an observational run.

The photos in Fig. 10.6 show various stages of the modification process.

To complete the telescope a data logger was built from a kit. The kit is still available from Magenta Electronics from their website, www.magenta2000.co.uk and it is KIT877. The design was published in EPE magazine in August and September of 1999. The heart of the logger is a PIC 16f877 microcontroller. The CPU comes already programmed with the software. The advantage of this kit is that it has eight channels and so can log data from multiple devices simultaneously. The disadvantage is that there is no real time clock on board, so careful note of the starting time is needed. It will happily run for hours on batteries, logging data to flash memory chips. There is not a great deal of memory, but with samples taken every 5–10 s it will record for a day. The data can be downloaded via serial cable to a PC (Fig. 10.7).

Observing Projects

Drift Scan Solar Transit

The Sun at Ku band frequencies has a diameter of half a degree, the same as the visual diameter. This means the radio emissions are emanating from the low regions of the solar atmosphere, just above the photosphere. At this frequency with this simple radiometer the solar output will appear to be fairly constant. Unlike UHF and VHF rapid changes in the solar output due to flares will not be visible.

Set up the dish fixed onto to something solid, such as a building wall or a heavy tripod, and aim it at an altitude where the Sun will drift through its beam. This is

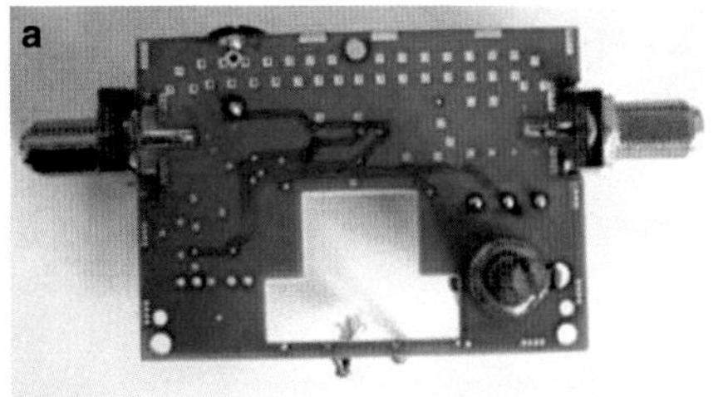

Fig. 10.6. (**a**) The circuit after removal of the meter looking from the front. (**b**) The view from the back showing the electronics; the buzzer is the black disc, and the potentiometer the silver disc. (**c**) This shows the removal of the RF connectors, the potentiometer, and the reversal of the buzzer. The wires for the potentiometer and the second wire for the buzzer have not yet been installed. (**d**) The finished receiver viewed from inside. Note on the left of the lid a small circuit board is soldered to the RF connector center pin containing a bias tee network. (**e**) The finished receiver viewed from the outside. The *top right hand corner* has some metal foil tape applied to cover a hole left by a redundant N connector. The case was "rescued" from a surplus sale, hence the engraved writing. The RF connectors are marked as suitable by accident! There is a 50 Ω terminator fitted to the output socket that is not being used at present. This should be a 75 Ω terminator in reality.

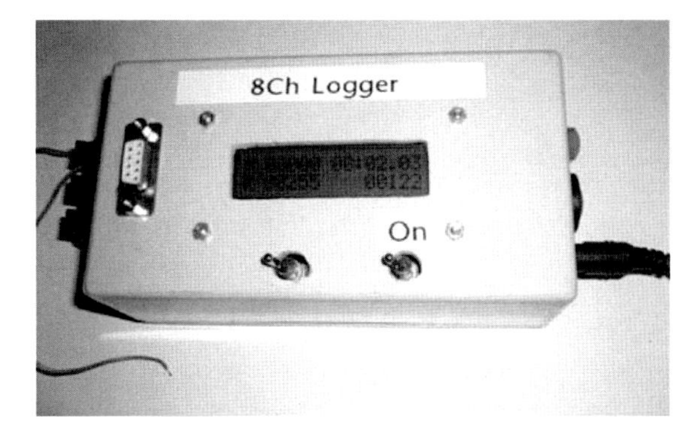

Fig. 10.7. Data logger, this one is based on a kit supplied by Magenta Electronics. There is more on data logging in a later chapter.

easier said than done with an offset dish. One method of aiming is to mount a flat metal bar about 150 mm long to the top edge of the dish so it overhangs at the rear. Determine the offset angle of the dish; this may be supplied with the dish, or it may be marked somewhere on it. It can be estimated from the angle between the center point and the center of the lnb feed, with respect to the normal at the center of the dish. Adjust the mounting bracket of the flat bar to the same offset angle so it effectively points in the direction the dish is looking, and ensure it is perpendicular to the rear face of the dish in the horizontal central position. Using a protractor with a plumb bob attached to its origin, hold the protractor against the underside of the bar; the plumb bob can be used to measure the altitude.

Allow the Sun to drift through the beam of the dish. The result can be used to measure the beamwidth of the antenna. Strictly speaking the beamwidth is measured between the 3 dB points, in other words, when the received power is half of the peak value. By measuring the time interval on the graph between the points where the signal is half its full value, the beamwidth can be estimated from the angular drift rate of the Sun (approximately 15°/h). For a Sky mini dish of 45 cm across, the beamwidth will be in the order of 4°.

The above argument relies on the assumption that the satellite signal strength meter is providing a measure of the received power and not something like the amplitude of the signal; after all, the system is completely uncalibrated.

Detecting the Moon

The Moon reflects solar energy towards us and is therefore theoretically observable. However the signal strength is much less than for the Sun. It is not easily detectable with a dish as small as 45 cm, so it may be a real challenge. You could try the same experiment with an 80 cm or 1 m dish. The signal will be close to the noise floor of the instrument.

Determination of the Equation of Time

By accurately setting up the dish to point due south, and maintaining its altitude to ensure the Sun always passes through the beam pattern, using the logger the time of solar transit can be determined. This will differ from midday clock time by the value of the equation of time. It will take a year of observations to achieve this, and the dish should be rigidly mounted in azimuth and smoothly adjustable in altitude.

The Slowly Varying Component of the Sun

At the operating frequency of around 12 GHz the instrument is able to observe the chromosphere of the Sun up a maximum altitude of 70,000 km above the photosphere. The radiation detected is largely thermal Bremsstrahlung from hot gases present in that region. The temperature of the gas varies over the solar cycle from a minimum at solar minimum to a higher level around solar maximum. The changes are slow and gradual rather than sharp and burst like. The enhancement in the solar output is due to the effects of sunspots.

At solar minimum the flux density observed is around 275 sfu (solar flux units), which is 275×10^{-22} Wm^{-2} Hz^{-1} so 1 sfu is 10,000 Jansky. Although no professional studies of the slowly varying solar output are being done now, the value of the solar flux density observed at solar minimum is considered to be constant enough that it is often used as a calibration source.

For a minute let's step back and look at what this output means for our telescope. Our lnb has a bandwidth of about 1 GHz, and a collecting area of 0.15 m^{-2}. The efficiency of the dish is estimated at 56%. Therefore the amount of power it receives from the quiet Sun is $275 \times 10^{-22} \times 1 \times 10^{9} \times 0.15 \times 0.56 = 2.3 \times 10^{-12}$ or 2 pW! Not only that, but the Sun gives a clear, strong signal, too.

Anyway, at solar maximum the overall output at 12 GHz will increase by a factor of about 1.37 over the value at solar minimum. A long-term project that would be interesting is to log the received signal power daily over the solar cycle to try and observe this slowly varying component. An observation period of 3–6 years should reveal a change. However it would be best if the instrument were calibrated against a known source to try and remove temperature effects from badly affecting the results. Don't forget if the dish is pointed at the Sun for extended periods the lnb could suffer significant heating, which will change the gain of the unit quite a lot. However alongside the previous experiment to determine the equation of time the transit instrument can be used to measure the solar flux density at noon by drift scanning. The Sun will only take 2 or 3 min to pass through the dish beam, so it will not cause overheating of the lnb, especially if the dish is a matte black color.

As for a calibration source use a compact fluorescent lamp (this is not an absolute calibration method – it is relative to the output of the lamp). Get one of the more powerful types around 26 W. These are usually constructed from coiled tubes. The lamp should be pointed into the dish beam end at a range of at least 0.6 m, but always the same distance each time. A power reading should be logged along with the Sun. The calibration measurement needs to be taken as close as possible in time to the solar reading so the lnb temperature is the same for both measurements. In practice you can't be there every day to take a calibration reading, but weekly measurements of solar flux would be sufficient to study the long-term changes. Note that the lamp should be retained for use only in this project and used for as short a duration as possible. These lamps age with use and their output changes. Also they take a short while to reach full output, so allow it stabilize for at least a few minutes before taking a reading.

When processing the results take the difference between the reading from the lamp and the reading from the Sun. This should remove receiver temperature effects sufficiently. In practice the brightness temperature of a compact fluorescent is going to be around 5,000 K and the Sun about 12,000 K.

A Low Cost Microwave Interferometer

This project is similar to the satellite finder telescope in that it uses low-cost satellite television components, but it contains some innovative trick concepts that make this one stand out. The idea was created by Dr. Alan E. E. Rogers of the MIT Haystack Observatory and is known as the VSRT (science loves its acronyms!), standing for the Very Small Radio Telescope. The project was funded by the U. S. National Science Foundation.

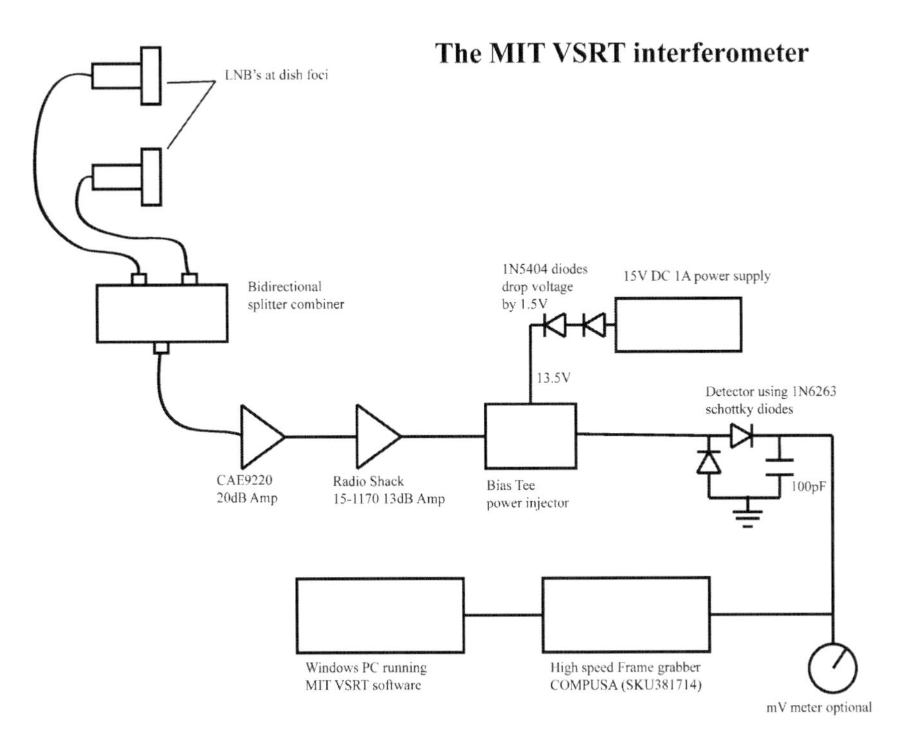

Fig. 10.8. The VSRT block diagram.

There are lots of notes, instructions, and downloadable software available from their website at http://www.haystack.mit.edu/edu/undergrad/VSRT/index.html. The memos section outlines a lot of the theory behind the instrument, which can be quite mathematical, but it is not necessary to understand all the theory to be able to build it, use it, and appreciate the basics of its function.

At the time of writing it's a new design and will probably be developed further, so keep an eye on the website. Figure 10.8 shows a block diagram of the telescope.

Let's look at the components in more detail. The LNB's are attached to dishes not shown in the diagram. In the United States small Direct TV dishes are suitable, in the UK and Europe Sky mini dishes are more readily available. Note, however, that Direct TV uses circular polarized LNB's and Sky mini dishes use linear polarized ones. But this does not affect the design or performance. Any dish, including larger ones, can be used with universal LNB's. Note that modern low-cost LNB's can have a significant drift in their local oscillators, which will affect overall performance. It may well be necessary to try several LNB's to find a good pair. Note also LNB feed horns for Sky mini dishes are not interchangeable with larger dishes that are more round in profile. The type of LNB should be matched to the type of dish used (ask your supplier if in doubt).

The LNB feeds are taken to a power splitter/combiner using patch leads with F connectors attached. The splitter combiner should have DC pass on all ports. This device is low cost and easily available on eBay, or via satellite TV specialists. A picture of one is shown in Fig. 10.9. The example illustrated shows a single in

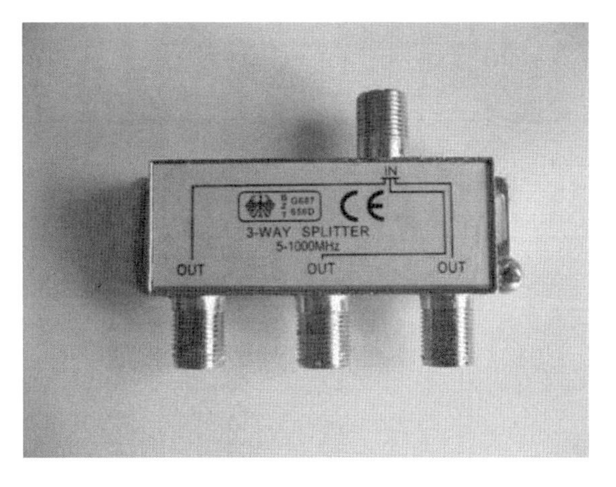

Fig. 10.9. Power splitter combiner.

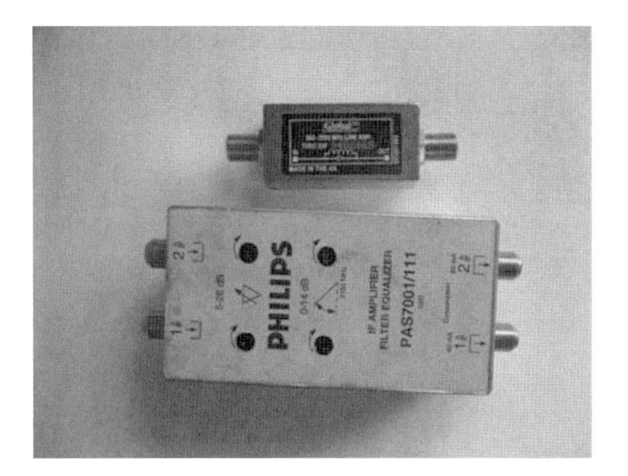

Fig. 10.10. VSRT in line amplifiers.

port, and three out ports, which are in fact reversible, and although not marked this one will pass DC power on all ports. You can confirm this with a multimeter set to measure continuity or resistance.

All ports on the splitter are matched to 75 Ω impedance, so all connecting feeders can be 75 Ω types. Use good quality double-shielded satellite TV feeders throughout.

Following the splitter the VSRT design calls for 40 dB of amplification, which is achieved using a pair of amplifiers in line on the output of the combiner (remember this is marked IN on the model illustrated because we are using it as a combiner rather than a splitter). A specific pair of amplifiers is detailed in the MIT documentation, but you can use a pair of surplus units you might have in your spares box. These are illustrated in Fig. 10.10. The larger amplifier is itself a dual unit, but you can use only one channel; this has variable gain of from 5 to 26 dB. The smaller amplifier has a gain of 14 dB. The combination of both units provides a range of amplification from 19 to 40 dB. Adjustability is not essential, but it is a

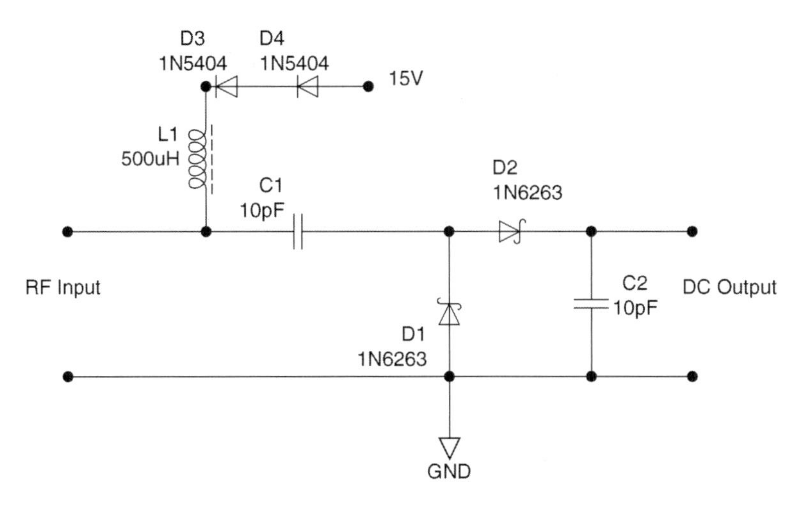

Fig. 10.11. Bias tee power injector (*left*), including a square law detector (*right*).

nice feature. The amplifiers need to be the wideband variety designed for cable, TV, and satellite TV use. If in doubt, obtain the models recommended in the MIT VSRT documentation. Some amplifiers will not work well if the input signal is too strong already (in a chain of amplifiers) and may oscillate, in which case you should try alternatives.

The next two modules in the chain are the only parts you need to construct, although you can buy the bias tee (sometimes called power injector). This method was used in the previous single-dish telescope. The circuit layout is given in Fig. 10.11, along with the circuit layout for the integrator. This integrator is of a square law configuration. The output is in proportion to the square of the input amplitude and is therefore proportional to the power of the input signal. This is only true if the device is driven with a signal of a suitable level. If the input signal is in the order of 30–250 mV it's about right.

The detector and power injector can be mounted in the same box. Use a die cast metal box as small as possible. If the box is small enough the components can be soldered directly to the connectors without the use of a circuit board. Try to keep the leads as short as possible. Use a power jack socket for the DC supply, a chassis F connector socket for the RF, and a phono socket for the DC output. Note the purpose of diodes D3 and D4 are to drop the voltage of the 15 V power supply to about 13.5 V. A 13.5 V power supply is not a common type, but a 15 V supply in the 1–2 A range will be easy to source. The diodes exhibit a voltage drop of about 0.7 V each, so the supply to the LNB's and amplifiers will be about 13.6 V. It is important to include these components because at 15 V it is not possible to know for sure what polarization mode the LNB's are in. At 13.5 V they are guaranteed to be the same. The chosen power supply should be of a regulated type. The diodes used in the square law detector are the Schottky variety, which have a low forward voltage drop.

The final hardware element is a video frame grabber, a particular USB-connected device manufactured by CompUSA, and its part number is SKU381714. This is the only one currently known to work with the software for the VSRT, distributed via the MIT Haystack website. The software package requires the Java runtime environment and has MS Windows drivers included geared especially for the CompUSA video capture device chipset. Follow their instructions for its installation.

How does the VSRT work? There are some clever ideas used in the design of the VSRT, which at first sight seem unlikely to work. The concept is quite straightforward. The LNB's are working in the 11–12 GHz frequency range, but the signals are internally down converted to a range of 1–2 GHz. These are then combined in the splitter/combiner, which is acting as a crude resistive mixer. The signals mix together. If the local oscillators of the LNB's were identical then the output of the combiner would be zero. But manufacturing tolerances mean the LNB local oscillators will probably differ by an amount up to 1 MHz. In fact we rely on the oscillators being different; in this way when the signals mix we get an output (the difference output) of up to 1 MHz. This is low enough to be manageable. The video frame grabber is used as a high-speed analog to digital interface, the second of the VSRT's clever design ideas. The bandwidth of a video capture unit is typically 4.5 MHz or better. The interference fringes generated and captured via USB are then integrated further in software and a power spectrum is displayed on screen.

What kind of observing projects are there for the VSRT?

Once again this instrument is a learning tool, its goal being to teach aspects of radio theory and design. It is capable of detecting the Sun, and it also could be used to record solar flux levels, but calibration is once again a major hurdle for its use as a scientific instrument. As in the previous receiver, relative calibration against a known source is a possibility. Again a compact fluorescent lamp of about 26 W could be used; however their microwave output can vary over a period of 10–20 s, and this output will change with lamp age and may vary between different lamps of the same type. Some experimentation will be needed to establish a suitable calibration. The comparison of the Sun signal with the lamp has got to be better than nothing at all. The team at Haystack are also working on calibration ideas for this project.

While observing the signal strength from the Sun changes in the Sun will be very hard to detect. The strength of the signal will depend on an amplifier gain, which itself will be influenced by ambient temperature. It will be affected by the variable attenuation in our Earth's atmosphere and will be dependent upon antenna pointing accuracy. This is also true for the single-dish telescope. This should not be an excuse to give up. Amplifier gain variation can be adjusted for by comparing signals with known sources. Not just the lamp method but the cold sky background at 12 GHz will show a figure about 8 K (3 K background plus a contribution from the atmosphere), although this will increase with lower altitude to about 35 K at 8° above the horizon. A hand placed in front of an LNB in "room temperature" conditions will show about 300 K. Take readings of all of these as a comparison along with the Sun. Record also the weather conditions at the time of observation, such things as air temperature, pressure, cloud cover, rain, etc. As for pointing accuracy mark the central point of the LNB's with a permanent felt marker pen. Attach a small piece of flat mirror to the dish face and carefully align it so that when it is pointing at the Sun the light reflection can be centered on the LNB. This offers the best chance of repeatability for each observation.

Radio signals from the Sun can twinkle like stars at night do in the visible band. This twinkling or scintillation is caused by the variable refractive properties in the radio band of our ionosphere. The VSRT could be used to investigate these phenomena.

A rather more advanced observing project, capable of resolving active regions on the Sun, is known as the closure phase method. This uses three dishes instead of two. The full description of this method is beyond the scope of this text, but you can read more about it in the memos section of the VSRT website.

Chapter 11

Building a Jupiter Radio Telescope

Here is one example of a radio telescope where you can buy a complete kit of parts to build yourself an instrument that will not only receive radio Jupiter but will also detect solar flares. It comes under the title of Radio Jove (more on this later). Referring to the Chapter 2 on the astrophysics of Jupiter it is clear the receiver needs to operate in the high HF radio spectrum. This means a conventional short wave receiver could be used as an alternative, but don't forget it's best to disable the automatic gain control (AGC).

As an introduction to the topic of receiving Jupiter let's look at what is required to turn a conventional radio with an audio output into a radio telescope with data logging functions.

First an antenna is required, whichever way you build the receiver. There are at least three variations used for this purpose. There is the good old wire dipole; you can refer to Chap. 5 for specific design parameters. Then there is the Radio Jove project, which recommends using a pair of dipoles, and lastly there is a loop antenna with a reflector. However there is some doubt about whether the loop version is very effective.

The best frequency to use for observing Jupiter lies in the range of from 18 to 22 MHz. This is sufficiently high to avoid the worst of the ionospheric cutoff problems (especially at night) and falls within a spectral band where Jupiter can get very noisy. The Radio Jove receiver is designed to work at 20.1 MHz. Of course this needs to be clear of other local noise or communications.

The Dipole

The first dipole recipe uses a trick to increase the bandwidth of a standard dipole antenna. By using three elements cut to three closely spaced wavelengths the antenna will be responsive to the three wavelengths simultaneously and reasonably good for the bits in between. A broader band dipole of this type would be suited to a telescope based on a tunable short wave receiver, where you could hunt for a frequency with the least interference. It is not worth building this for a fixed frequency instrument such as the Jove receiver. Note when combining more than one

J. Lashley, *The Radio Sky and How to Observe It*, Astronomers' Observing Guides,
DOI 10.1007/978-1-4419-0883-4_11,

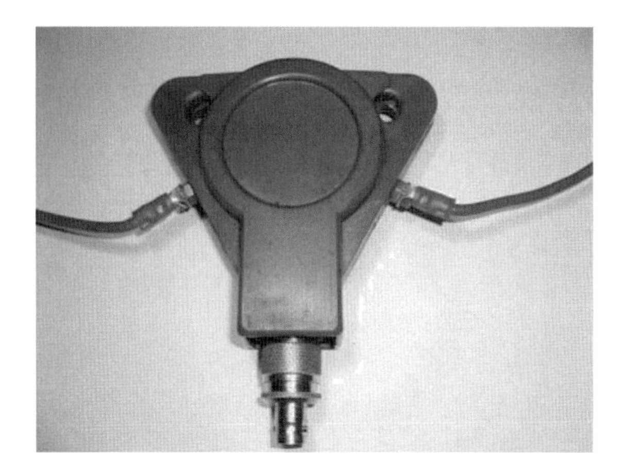

Fig. 11.1. Dipole antenna. This one is based on a commercially made balun. There is an adapter fitted to the base to provide a BNC output. The wire used was a 2.5 mm^2 electrical mains cable. Additional elements can be fitted to increase the bandwidth so long as they are not mutually resonant. For an antenna to work between 18 and 24 MHz use three pairs of wires a total length of 7.94, 6.81, and 5.96 m wide. See Chap. 5 for more information.

dipole into the same aerial using a common feed point it is important that design wavelengths are not integer multiples of each other, so that only one set of dipoles can be in resonance at any time.

Fig. 11.1 shows a drawing of the antenna with a picture of the 4:1 balun that was purchased from an amateur radio equipment supplier. The wire lengths are given in the figure.

The aerial should be suspended in an east–west orientation, so that it is most sensitive in the north–south direction.

The Dipole Array

A single dipole antenna is most sensitive in a perpendicular direction to its orientation, but this sensitivity is theoretically distributed 360° around the antenna, although in practice the ground influences the beam pattern. At the time of writing Jupiter is low in the sky, about as low as it can get, in fact. This is worsened the further north you live. So the aim of the dipole array is to manipulate the beam of the antenna to increase its effectiveness in a given direction, to maximize our potential to observe Jupiter.

The Jove project sells a complete kit of parts to build a dipole array, but it is a simple thing to construct from locally sourced materials. The kit contains plastic tubing, the sort used for household plumbing, to construct masts to support the antenna. Two pairs of masts are needed. However the lower the antenna beam the higher above ground the dipole height must be increased. If Jupiter is relatively high in the sky, then a height of 3 m is OK, but when Jupiter is low as it was in 2009 a height of 6 m is better. Plastic tubing is a bit flexible in lengths a few meters long, but can be supported by attaching guy ropes at several points and tying these to ground stakes. Eye bolts can be inserted through the pipe to attach the guy ropes. To allow for a variable height adjustment, use a larger diameter pipe for the first 3 m, and a smaller sliding fit pipe for the upper 3 m. Guy ropes will be needed at least at the top of the

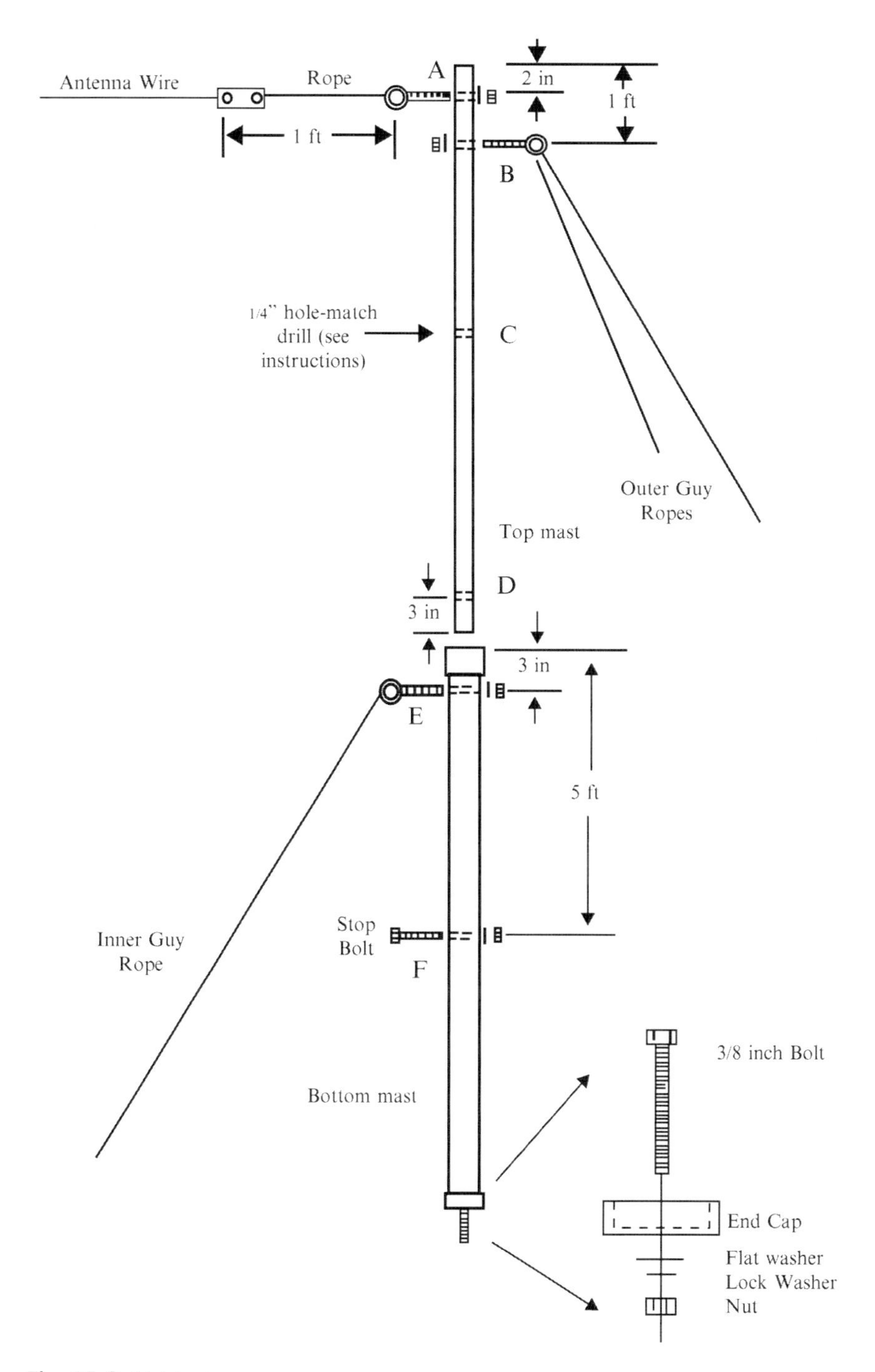

Fig. 11.2. A typical mast.

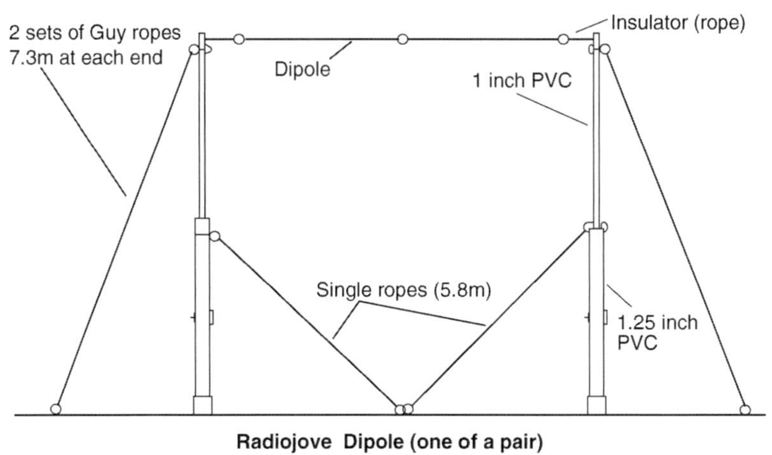

Fig. 11.3. Side view of the dipole and mast construction.

base section and the top of the moving section; more may be needed to improve rigidity. Alternatively steel or aluminum pipe could be used, such as scaffold poles or large conduit, but guy ropes are still recommended (Figs. 11.2–11.5).

The Jove antenna uses a combination of a phasing cable and height above ground to control the beam pattern altitude. The layout of both dipoles is in an east–west direction separated by 6 m (20 ft). The extra 0.375λ phasing cable is adding into the path of the southern dipole before the two dipoles are combined. Don't forget the velocity factor of the coaxial cable. RG59/U has a velocity factor of 0.66 so the length of the phasing cable is 0.375 × 14.93 × 0.66 m, which is 3.69 m (assuming a Jove frequency of 20.1 MHz). This layout schematic is for middle to high Northern hemisphere observers.

Table 11.1 lists the height above ground to mount the aerials for a given altitude of Jupiter.

The power combiner could be easily made, but in fact it is so cheap to buy it would cost you more in RF connectors and boxes than it is to buy one. Power combiners are marketed for television installations, allowing two televisions to use a common aerial. They take care of impedance matching. Note that the ones you want are splitter combiners. They work both ways to combine signals together or to split a single signal into two. They usually come with F connector sockets and operate over a band of 5–2,400 MHz. F connectors can easily be fitted to RG59/U cables. Seal all cable connections and joints with self amalgamating tape to keep water from getting into the cable insulation.

The Loop Aerial

This type of antenna was written about in a few journals some time ago. It consists of a broken loop 1.37 m in diameter made from small-diameter copper tubing. Something like car brake pipe material would work. This is supported on insulating

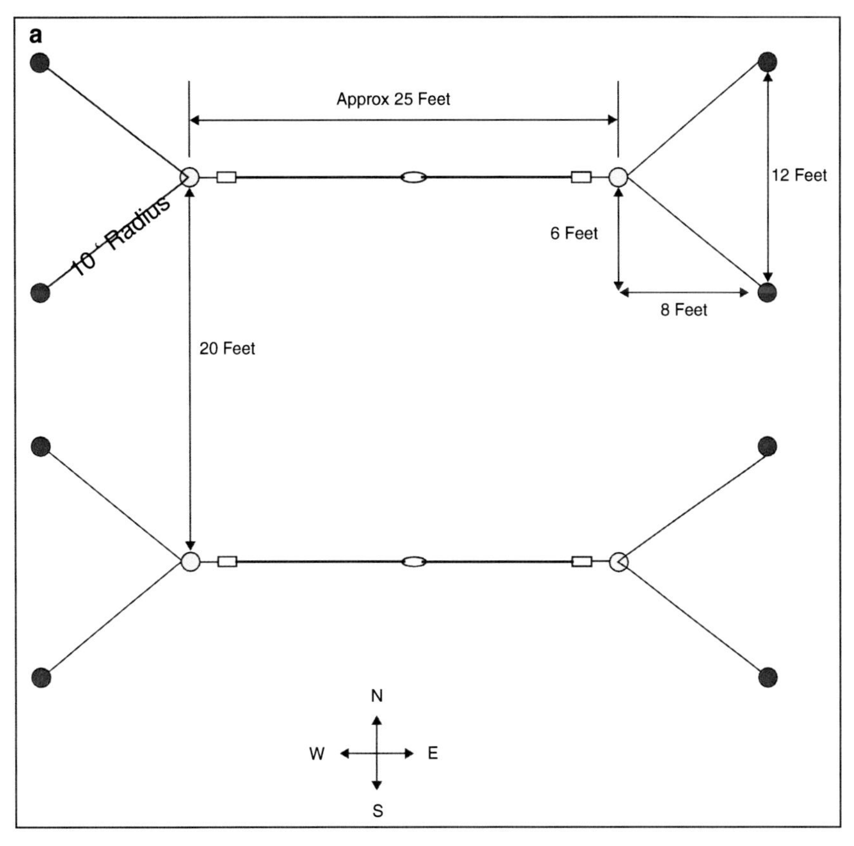

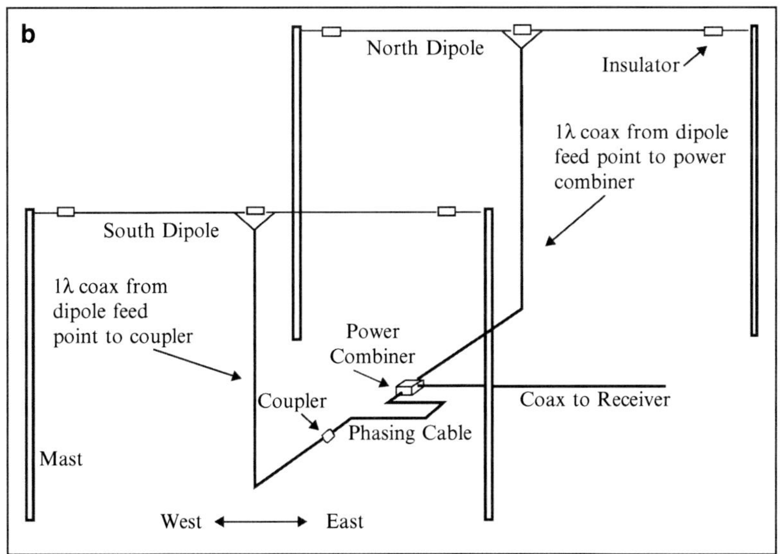

Fig. 11.4. (**a**) The birds eye view of the dipole array layout. (**b**) Another view of the dipole array showing the feeder and phasing harness.

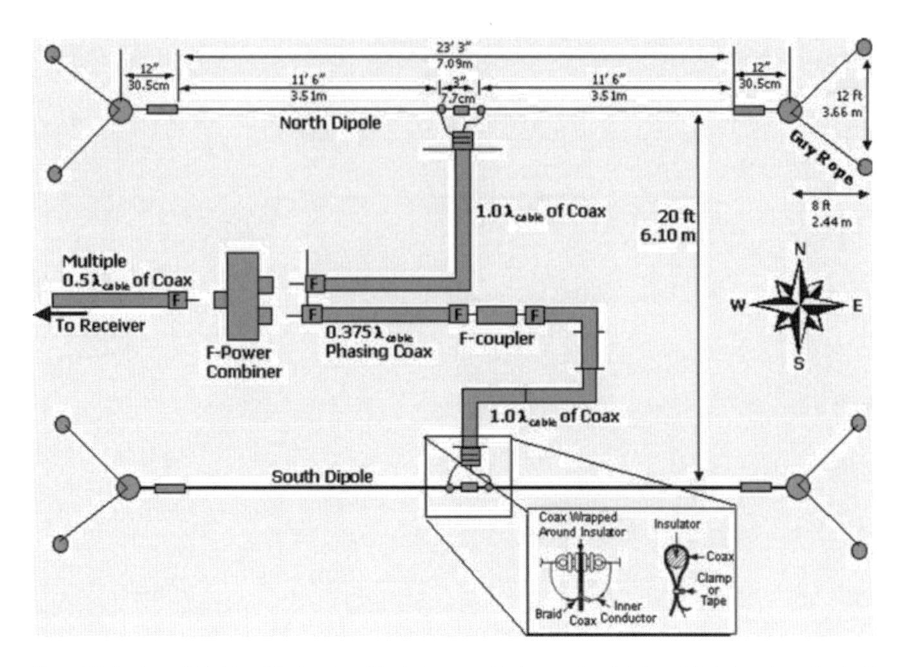

Fig. 11.5. A more detailed view of the dimensions of the dipole array, and feeder connections. Note the main feeder to receiver section should be in half wavelength multiples.

Table 11.1. Antenna height above ground for different Jupiter altitudes

Jupiter elevation (°)	Aerial height (m)
20–40	6.1
40–55	4.5
55–70	3.0

stand offs from a square frame 1.8 m across at a height of 17.8 cm. The frame is covered with chicken wire mesh to act as a reflector. Coaxial cable is connected between the open ends of the loop and fed to the receiver or preamplifier. There is doubt about how effective this is, so here is a challenge for you. As an observational experiment test the loop antenna alongside the dipole one to determine how suitable it is. It would be best to build two receivers, though, so side by side comparisons could be made under the same observational conditions.

The Receiver

Many radio astronomy observers use the Icom IC-R75 receiver. This is certainly a capable receiver, with the distinct advantage of being able to turn off the AGC control. While at the time of writing this radio is listed on Icom America website, it does not appear to be officially available in the UK, for some reason. Other Icom products that are suitable are the computer-controlled receivers such as the PCR1000. The PCR1000 is no longer available new, but it had the AGC off func-

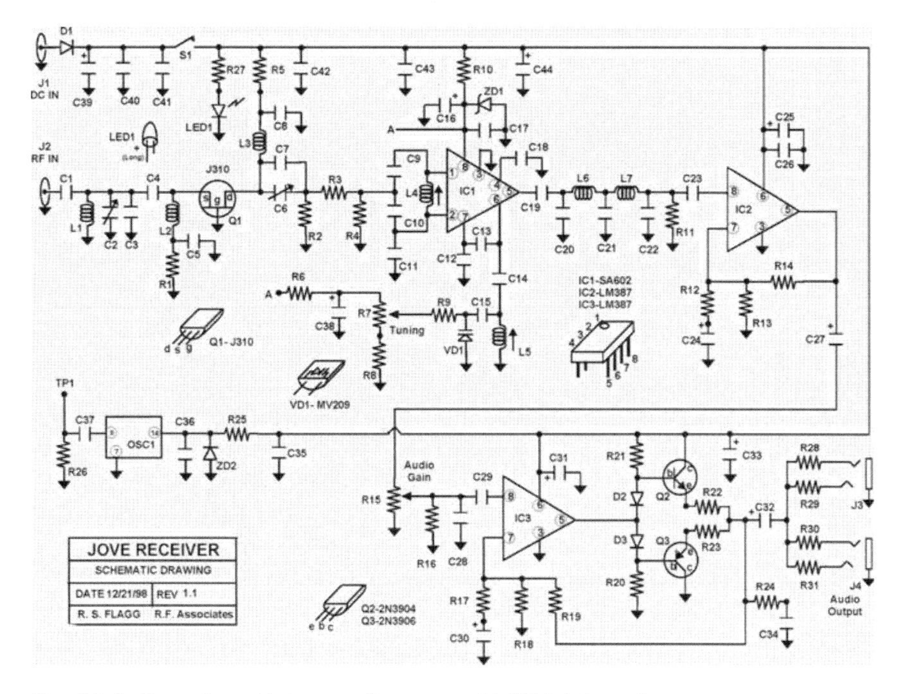

Fig. 11.6. Schematic drawing of the Jove receiver. (Diagram courtesy of the NASA Radio Jove team).

tion that is desired. The manuals for the PCR1500 and PCR2500 models do not list this feature. They indicate only AGC slow and fast – this is not the same as off. Unfortunately the AGC off control is not required much by amateur radio enthusiasts, so even the hardware dealers are not always familiar enough with their products to advise on it. It is always advisable to consult the manufacturers manual before you buy.

If you have not already got a suitable communications receiver, it would be cheaper to buy the Radio Jove kit, purposely designed for the job and a lot more fun knowing you built another radio telescope yourself. The manual for the receiver and antenna kits can be downloaded from the home page http://radiojove.gsfc.nasa.gov/, although you may have to trawl through a few of the pages to find it. The circuit diagram and PCB layouts are included in the book, so theoretically you could construct it from locally sourced parts, but this would only save a few dollars on shipping costs (Figs. 11.6 and 11.7).

The output of the Jove receiver is audio, which you could listen to by attaching a speaker. Sometimes Jupiter sounds like waves rolling up a pebble beach. More practically, however, it is designed to be logged using the Radio Skypipe software available from http://www.radiosky.com. See Chap. 13 for more information on this software.

One thing we have not covered yet. If you are going to use an existing radio with audio output, or even the Jove receiver, but you don't want to run a PC all day every day, you can easily build an integrator circuit to convert the audio output to a DC voltage, which can be logged to flash memory with a data logger. The circuit diagram is shown in Fig. 11.8.

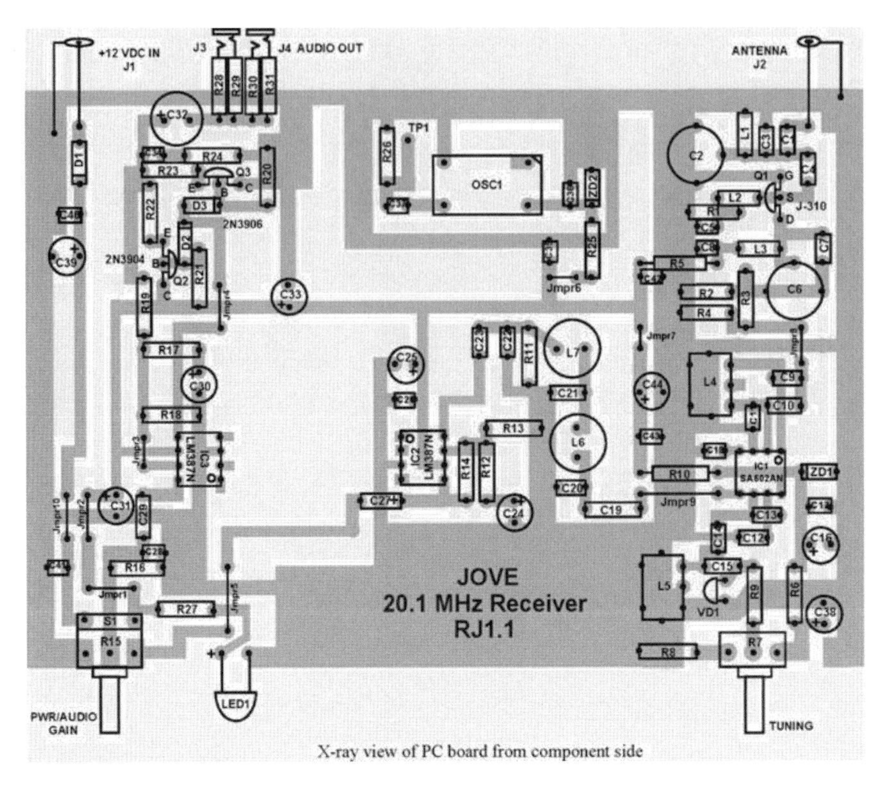

Fig. 11.7. X-ray view. (Diagram courtesy of the NASA Radio Jove team).

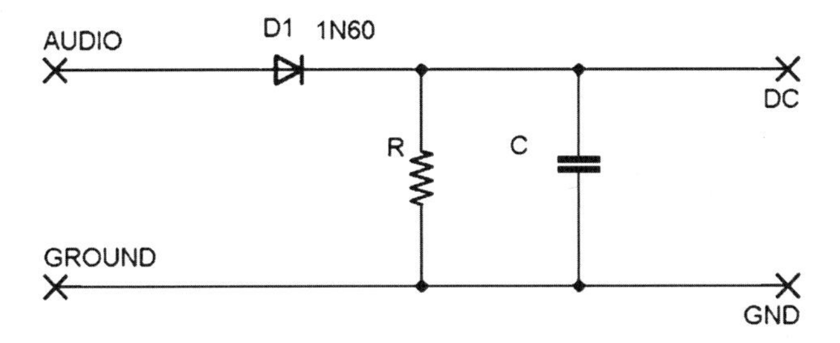

Fig. 11.8. Audio signal integrator.

The diode rectifies the audio signal from the output of the radio, and R and C together form the integrator. The choice of R and C defines the time constant of the circuit. The function of the integrator not only smoothes but provides a DC output that can be logged. The choice of time constant will vary, depending on the type of observing you wish to do. The down side of the integrator is it will mask very fast changes. Time constants of between 0.1 and 5 s would be a reasonable range. With modification and the use of a rotary switch the integrator could be fitted with several sets of RC values. The formula for calculating time constant t is

$$t = RC$$

where R is in ohms and C is in farads.

To calculate a given time constant, start by assigning a resistance value, and calculate the value of C. If C is, for example, unusually high in value, increase the resistance and try again. For example if you chose $R = 100\ \Omega$ and a time constant of 1 s, C would have to be 0.01 farads, rather a large capacitor. By choosing a value for R of 100 K then C is a more reasonable 10 μF. A value like this will be an electrolytic type that is polarized. So it needs to be placed in circuit the correct way around with the negative pin to ground. The diode shown is a germanium type, chosen because it has a small forward voltage drop of about 0.2 V. Silicon diodes drop about 0.7 V and should be avoided.

Chapter 12

Building a Broad Band Solar Radio Telescope

This is the most complex of the projects presented in this book. However it offers a lot of potential to learn about how receivers work, and as a useful radiometer or even spectrometer for solar flare observation.

Construction of ultra-high frequency circuits is not really suitable for a beginner. However, some of the difficulties usually encountered have been removed by the use of a readymade tuner module that translates the frequencies in the range 51–853 MHz down to a workable 37 MHz. The remaining components required to turn it into a radio telescope you can construct.

The type of tuning module used in this project is commonly found in television receivers, cable TV boxes, and other similar television tuner applications. Suitable units can be removed from used equipment, although they are inexpensive to purchase new if you can find a supplier willing to sell a single unit.

Background

The approach taken here is to break down the radio into its basic modular units. Each module will be treated as a separate entity and built on its own board. The modules are then shielded where necessary and installed into an RF screened box. There are two main reasons for this approach. First of all, each unit can be built and tested separately, so that when it comes to putting it all together, troubleshooting problems will be easier. Secondly, the resulting instrument would be easy to modify later by replacing individual modules, or adding more. For example, the frequency band it operates on can be changed by replacing the front end RF section.

Before we begin to design the instrument, let's outline our goals:

1. The instrument should be capable of measuring the strength of a fixed frequency channel, covering a reasonably wide but not excessive bandwidth. In other words, it should operate as a total power radiometer telescope.
2. It should be able to scan a range of frequencies in a predetermined time interval and to measure the changes in power levels across the band of frequencies. In other words, it should be able to act as a spectrometer.

J. Lashley, *The Radio Sky and How to Observe It*, Astronomers' Observing Guides,
DOI 10.1007/978-1-4419-0883-4_12,

3. It should be possible to automate the control of the radio functions, so it can be left unattended to do its work.
4. The output of the receiver should be in a form easy to log using either a PC or a stand-alone data logger.
5. The frequency of coverage should be a wide as possible.

These tasks may seem daunting. However, following these plans is not that hard, so long as you have patience to persevere when things are not working out right.

The design was inspired by the E-Callisto project, a low-cost solar flare spectrometer designed by the team at ETH Zurich. You may wish to build that instrument as an alternative. The plans may be found at: http://helene.ethz.ch/instrument/callisto/ecallisto/applidocs.htm.

In Chap. 7, you saw the block diagram of a superheterodyne radio receiver. That is what we are about to design here. The basic superheterodyne will have a single intermediate frequency section, but it can have more. This design has two intermediate frequencies, and is therefore called a double conversion superheterodyne.

Design Overview

The heart of the receiver is a television or multimedia tuner module. This avoids the difficulty of constructing stable high frequency circuits. The television tuner chosen for this project is a CD1316L, manufactured by Philips for the European market. There is a similar model, the CD1336L, designed for the U.S. market, but this project implies the use of the CD1316L because the IF outputs differ on the two models, and the frequency coverage is slightly different. While we are on the subject of tuners we should explain there are many different variations out there.

Older tuners used an analog 2–30 V signal to select the frequency of operation. The CD1316L uses a digital data line to set the operating conditions. The control bus is known as I^2C. Most modern tuners use this technique now. Those tuners aimed at the terrestrial television market usually have gaps in their range, because television channel allocations are broken up into several bands separated by bands used for other purposes. The CD1316L is more commonly used for cable television boxes or similar multimedia units, and therefore has continuous frequency coverage, although the range is broken into three switchable bands. The design could easily be modified to use any available I^2C tuner, but it would be important to obtain the data sheet to identify any changes that need to be made such as tuning calculations, and to see what frequency ranges are available. Any of the tuner modules could be used as a UHF scanner from about 470–860 MHz.

There is a third flavor of tuner that is less useful for our purposes. This includes demodulators on board providing a video and audio output. This type is sometimes found in computer-based television cards. This telescope design requires access to the intermediate frequency output before any demodulation can occur.

The output of the tuner, the first IF, is a fixed frequency band 7–8 MHz wide in the range 32–39 MHz. Since here we are proposing a much narrower bandwidth than that for back end processing we can select a suitable frequency of 37.7 MHz as our first intermediate frequency with a bandwidth of, say, 330 kHz. By adding switchable filter banks we could offer a selection of bandwidths to suit the application.

The first intermediate frequency is passed to a mixer stage of our construction. The reason a first IF of 37.7 MHz is suggested is that by mixing this with an oscillator signal at 27 MHz, a convenient second IF of 10.7 MHz is achieved. This is a common IF frequency used in television circuits, and suitable ceramic filters are easily obtained. An overtone crystal of 27,000 MHz is also readily available to construct the down converter.

Finally, the second IF can be further amplified and passed to a detector stage, in this case a logarithmic amplifier, the output of which is a DC voltage proportional to the signal power and is easily captured to a PC with an ADC (analog to digital converter).

Construction

The front end in this case is the TV tuner module, which is designed to simply connect directly to an antenna. There is a variable gain amplifier built into the unit that is controllable via a DC voltage that we can easily make manually adjustable using a potentiometer.

Figure 12.1 shows the receiver prototype, and Table 12.1 identifies the pin functions.

Pin number begins next to the RF input sockets. The only pins needed here are 2, 5, 7, 8, 9, 11, 14, and 15. The IF output pins provide a balanced output from pin 14 and 15, or you can use an unbalanced output from either pin 14 or 15 and ground. The ground connections are four tags on the four corners of the base that form part of the case.

The I^2C needs some introduction. It is a databus technology introduced by Philips and is found in a number of modern IC devices such as EEprom memory chips. It defines the way in which communications of digital data are handled and is a form of serial interface bus. The data is sent via a two-wire system. The lines are identified by the SCL and SDA references. SCL refers to the "clock" line and SDA

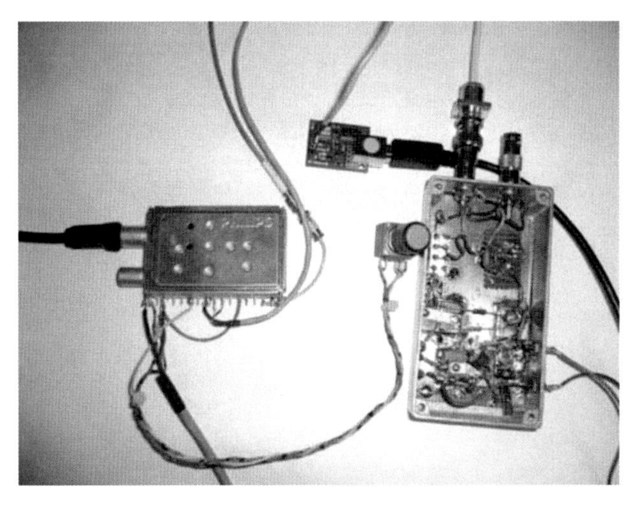

Fig. 12.1. To the left is the tuner module, to the right is the home built mixer oscillator. The small board upper centre is a USB to L2C interface.

Table 12.1. Tuner module pin numbers and their function

Pin number (from the antenna socket)	Function
1	No connection
2	+5 V DC at 80 mA max
3	No connection
4	No connection
5	RF gain control 0.4–3.3 V DC
6	No connection
7	I^2C address select
8	I^2C SCL
9	I^2C SDA
10	4 MHz reference output – not required in this design
11	+5 V DC at 110 mA max
12	ADC input – not required in this design
13	No connection
14	IF output
15	IF output

Table 12.2. Five bytes of data are required to program the tuner - see text

Name	Byte	Bits (most significant bit first)							
Address	ADB	1	1	0	0	0	CA1	CA0	R/W
Divider byte 1	DB1	0	N14	N13	N12	N11	N10	N9	N8
Divider byte 2	DB2	N7	N6	N5	N4	N3	N2	N1	NO
Control byte	CB	1	CP	0	T1	TO	1	1	0
Band select	BB	0	0	0	0	0	P2	P1	PO

Table 12.3. The voltage on pin 7 defines the L2C address

CA1	CA0	Voltage applied to 7 (V)
0	0	0
0	1	Terminal open
1	0	2–3
1	1	4.5–5

to the "data" line. Pin 7 is not really part of a standard I^2C bus, but it is used in this device to define the address to which the bus will respond. The tuner can only be set to one of four addresses. Table 12.2 illustrates its function.

To set up the tuner, the address and four data bytes are sent to it. The bytes are made up from binary bits, but are sent to the tuner in hex format.

Since we're only using a single tuner, it is easiest to ground pin 7 to the 0 V power line; this defines the address as binary 1100000(*). Bits 2 and 3 (CA0 and CA1) of the ADB byte are then 0. The last bit tells the tuner whether we want to read or write data to it. The R/W bit is 0 when writing data to it, so the first byte becomes 11000000, which in Hex is C0. This will not change. See Table 12.3 for addressing options.

The second and third bytes DB1 and DB2 define the frequency of operation. A single step relates to a change of 62.5 kHz in tuning frequency. The following formulae is used to calculate the byte values for a given operating frequency.

Table 12.4. Bits P0, P1 and P2 of the last byte BB define which of the three bands the tuner will operate.

Band	Minimum frequency (MHz)	Maximum frequency (MHz)	P2	P1	P0
Low	51	171	0	0	1
Mid	178	450	0	1	0
High	458	858	1	0	0

$$N = \frac{f_{in} + f_{if}}{62500}$$

where f_{in} is the input frequency and f_{if} is the required IF (37.7×10^6 in our case).

The resulting N needs to be rounded to the nearest whole number and expressed as a hexadecimal value. The minimum value for N is 1419 decimal or 58B in hex where DB1 = 05 and DB2 = 8B. Do not set the frequency to lower values than this, and do not try to exceed the upper value representing 858 MHz.

The control byte should have T1 and T0 set to 0. CP sets the charge pump value. For fast tuning CP should be set to 1, and if the frequency is going to remain fixed for a while, should be changed back to 0. In spectrometer mode the binary value of CB is 11000110 or C6. For the radiometer user it should be 10000110 or 86.

The final byte of data is the frequency band selection. Table 12.4 shows how this is broken up into three parts.

To set low band the byte BB is 00000001 or hex 01, mid band is hex 02, and high band is hex 04.

Now let's test the tuner. For this you will need a spectrum analyzer. The best value analyzer to start with is a computer-based USB oscilloscope adapter with an FFT function. One model from China, that is sometimes available on eBay is a DSO 2150. The bandwidth is a decent 60 MHz; this makes it very useful for testing and analyzing receiver IF circuits. Other models are available with up to a 200 MHz bandwidth.

In order to actually send the data to the tuner we need some form of interface hardware. This is usually done with a microcontroller. A microcontroller is a form of single chip computer. If the bug really bites, and you get involved in more hardware development, it is recommended you take the time to learn about using and programming microcontrollers, such as the range available from Microchip known as PIC's, or the range available from Atmel. However, a whole book could be dedicated to this task alone, and many are available. So to ease that extra burden we recommend experimenting with a serial to I^2C or a USB to I^2C interface, and use your PC to control the receiver. You can purchase a BV4221 for development testing from a company called ByVac Electronics (see http://www.byvac.com). The drivers supplied with it install it as a virtual COM port on the PC, and a simple terminal emulator was supplied on a CD called "bv_com_a.exe," which is highly recommended for this job.

Now connect three wires from the I^2C adapter, the ground wire to the tuner case, and the SCL and SDA lines to pins 8 and 9, respectively.

Next connect the positive lead from a 5 V regulated power supply to pins 2 and 11, and the power ground to the case of the tuner.

Normally the receiver gain is controlled automatically by external circuitry. However we have already seen that we need to manually control the gain. For this purpose a 0–3.3 V variable voltage is supplied to pin 5. This needs to be derived from the 5 V supply using a potential divider (refer to the Chap. 8). Use a potentiometer to vary the voltage, along with a fixed resistor to limit the maximum voltage to 3.3 V. For example if you chose a 1K potentiometer R2, the fixed resistor R1 will be

$$R1 = \left(\frac{R2.V_{in}}{V_{out}} \right) - R2$$

This gives

$$R1 = \left(\frac{1000x5}{3.3} \right) - 1000$$

This means R1 is 515 Ω. Chose the nearest fixed resistor value higher than the calculated value which is 560 Ω in standard ranges. Connect the fixed resistor between the 5 V supply and one end of the potentiometer, and connect the other end of the potentiometer to ground. The variable wiper output is then connected to pin 5 of the tuner.

Check all connections carefully. Then connect the BC4221 adapter to the PC and open the terminal software; set the speed to 115,200 baud and the COM port installed by its driver and click on the button to the left of the COM port box. The icon will turn from red to green. Press the enter key and you should see a hex address appear. Apply 5 V power to the tuner, and connect an FM radio antenna (in practice a piece of wire is probably sufficient for testing).

To test the tuner chose a strong FM broadcast channel and determine the divider bytes DB1 and DB2 for that channel. You might choose a channel called Smooth Radio at 106.6 MHz, a strong FM radio channel in the UK.

ADB = C0

DB1 and DB2 are $(106.6\times 10^6 + 37.7\times 10^6)/62500$ = decimal 2308, so DB1 = 09 and DB2 = 04 hex

CB = C6

BB = 01

In the terminal software window type the following string: s-c0 09 04 c6 01 p and press the enter key. The “s” command tells the adapter to be a master controller and open communication to address c0. The four data bytes follow and the p closes communication. The tuner should now be locked onto a frequency 106.6 MHz converting it to 37.7 MHz at the IF pins 14 and 15. To check it is working, connect channel 1 of the oscilloscope to pin 14 and ground and turn on the FFT. Adjust the settings so you see 37 MHz roughly central in the FFT window. You should see a peak at around 37.7 MHz, which is the radio transmission. Alternatively you can use an RF signal generator and try it on other frequencies. Adjust the RF generator frequency to see the peak move sideways in an 8 MHz window (the bandwidth of the tuner output). The spectrum from my tuner is shown in Fig. 12.2

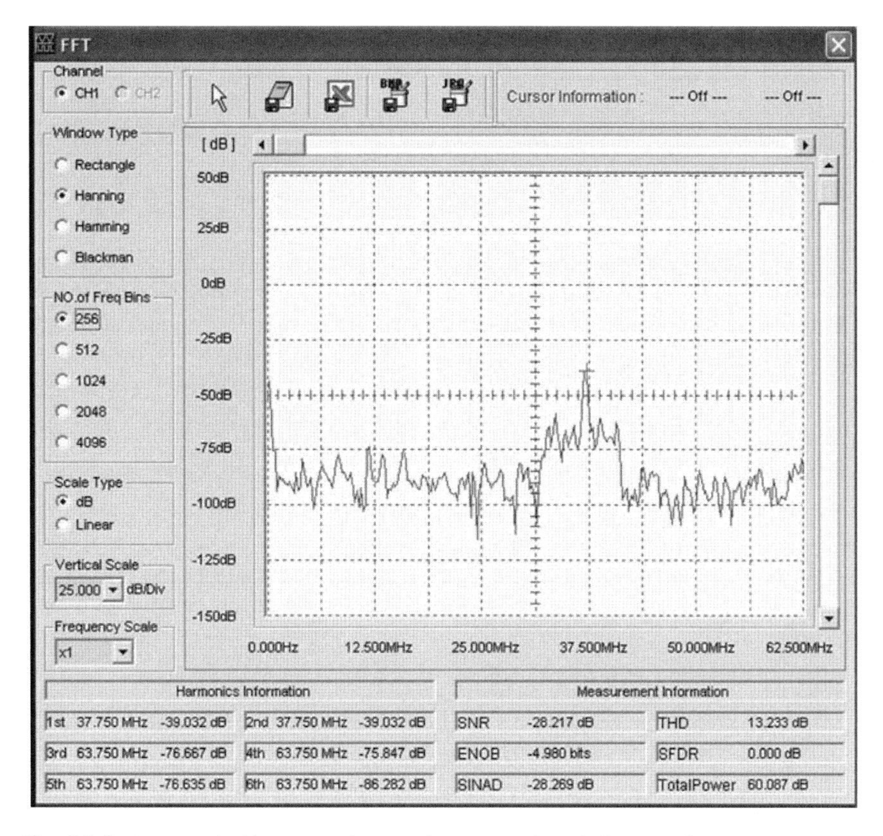

Fig. 12.2. FFT spectrum plot of the tuner output showing a peak at 37.75 MHz. The signal is from an FM radio station at 106.6 MHz.

Table 12.5. When reading back information from the tuner only 2 bytes are needed.

Name	Byte	Bits (most significant first)							
Address	ADB	1	1	0	0	0	CA2	CA1	R/W = 1
Status	SB	POR	FL	1	1	0	AD2	AD1	AD0

Finally you can read data from the tuner, but only one byte is returned. However there is no way of confirming what frequency it is tuned to. Table 12.5 shows the format of the Status byte.

When the address sent to the tuner has a 1 added in the least significant bit position, it tells the tuner to return the Status byte. The POR flag is set to 1 if the tuner has not been configured since a power on reset; otherwise it is 0. The FL flag returns a 1 if the phase locked loop inside the tuner is locked. The AD2, AD1, and AD0 bits return the value of the internal analog-to-digital converter. Mostly there is little value in reading the Status byte, although sometimes it would be useful to confirm that the tuner has locked after a request to tune a channel.

The Second IF Stage

The IF output of the tuner can be treated as either a balanced or an unbalanced output. In this project we will use the single output – it does not matter which one. It is always a good practice after a mixer stage to use a filter to remove unwanted products. These TV tuners are designed to drive a SAW filter (surface acoustic wave). But you can try making your own. Filter design is covered later.

The next stage is a frequency mixer, which will reduce the first intermediate frequency to 10.7 MHz. This means an ordinary HF communications receiver could be used to help set it up and test it. It also means useful ceramic filters are easily obtainable for 10.7 MHz, in a wide variety of bandwidths.

A mixer circuit is a fundamental building block of nearly all radios. So this part of the construction process is a useful learning exercise. The mixer itself is a readymade unit by Minicircuits, the part number being SBL-1, and can be ordered online direct from Minicircuits or from other RF component stock lists. A mixer combines the RF signal with a local oscillator, to produce two outputs – the sum of the two inputs and the difference of the two inputs. The SBL-1 is an example of a double balanced mixer, meaning the original radio frequency and the local oscillator frequency are strongly suppressed and don't form a part of the output to any significant level.

The schematic of the oscillator is shown in Fig. 12.3; it is a Colpitts type.

Note here that the Crystal X1 merely controls the frequency. The inductor L1 and capacitor C1 form a series tuned circuit that also resonates at the desired frequency. If the crystal is removed, the oscillator should still work at approximately the correct frequency, but it will be a lot less stable and much more temperature sensitive. The values of L1 and C1 are calculated to resonate at 27 MHz (see Chap. 8).

The first transistor is essentially an amplifier. However some of its output is fed back via C2 to the base pin and is re-amplified. The amplifier is set up to be unstable, therefore creating the oscillations we desire. The output of the oscillator is fed via C5 to a buffer amplifier, to isolate the oscillator from the mixer to avoid loading problems.

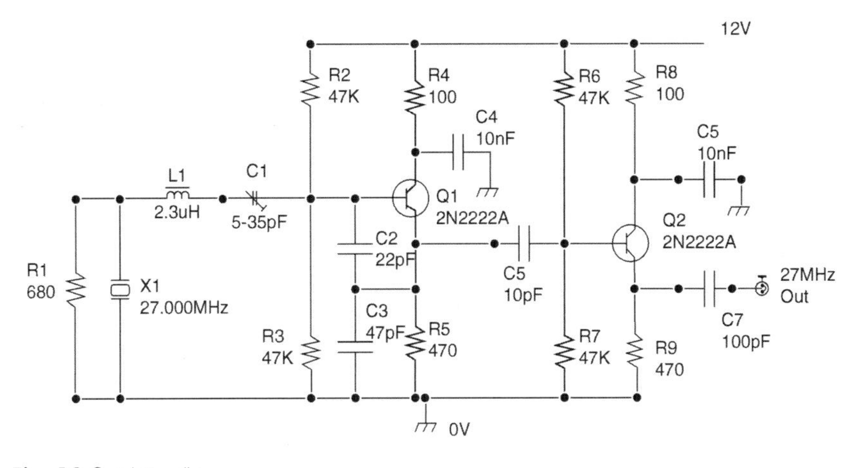

Fig. 12.3. Colpitts oscillator.

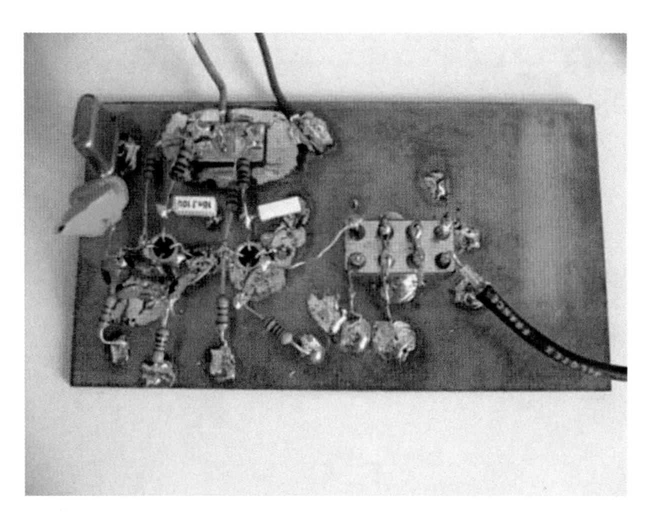

Fig. 12.4. A mixer built dead bug style.

Note the series tuned circuit formed by L1 and C1 is not only resonant at 27 MHz, but also presents low impedance at resonance. Together with R1 this damps out the chances of the crystal resonating on its fundamental frequency, and forces it to resonate at the chosen design overtone. When building the oscillator the crystal can be removed and L1 and C1 tweaked to produce near to the design frequency. A digital frequency counter or an oscilloscope is needed for this task. Once set up return the crystal to its position and the counter should confirm a solid frequency lock to 27,000 MHz (Fig. 12.4).

The mixer used is a Minicircuits SBL-1+. Its pin configuration is shown in Fig. 12.5.

The output of the oscillator is connected to pin 8. The oscillator ground is connected to pins 2, 5, 6, and 7. The output from the tuner is connected between pin 1 (ringed in blue) and ground. Pins 3 and 4 are connected together. Finally the output is taken from the junction of pins 3 and 4 to an RF connector or to the next stage where the body of the connector is also grounded.

This author's prototype mixer was constructed dead-bug style. A photograph is shown in Fig. 12.4. You could, of course, etch a circuit board instead for a more professional look. I would suggest you try both examples of construction. Build one module ugly style and another by etching a board as part of a learning process and then decide for yourself which you prefer. The board layout is in Fig. 12.6.

The holes for drilling for drilling are shown as circles, attach the overlay drawing to the top of the board with tape, and drill the component holes carefully. Most holes will be about 0.6 mm in diameter, but the trimmer capacitor and pad holes will be more like 1 mm.

X2 is the IF output, X1 is the RF input, refer to Fig. 12.6 text for power connections. The board should be mounted into a metal box, preferably a die cast type ensuring the box is well grounded to the circuit ground.

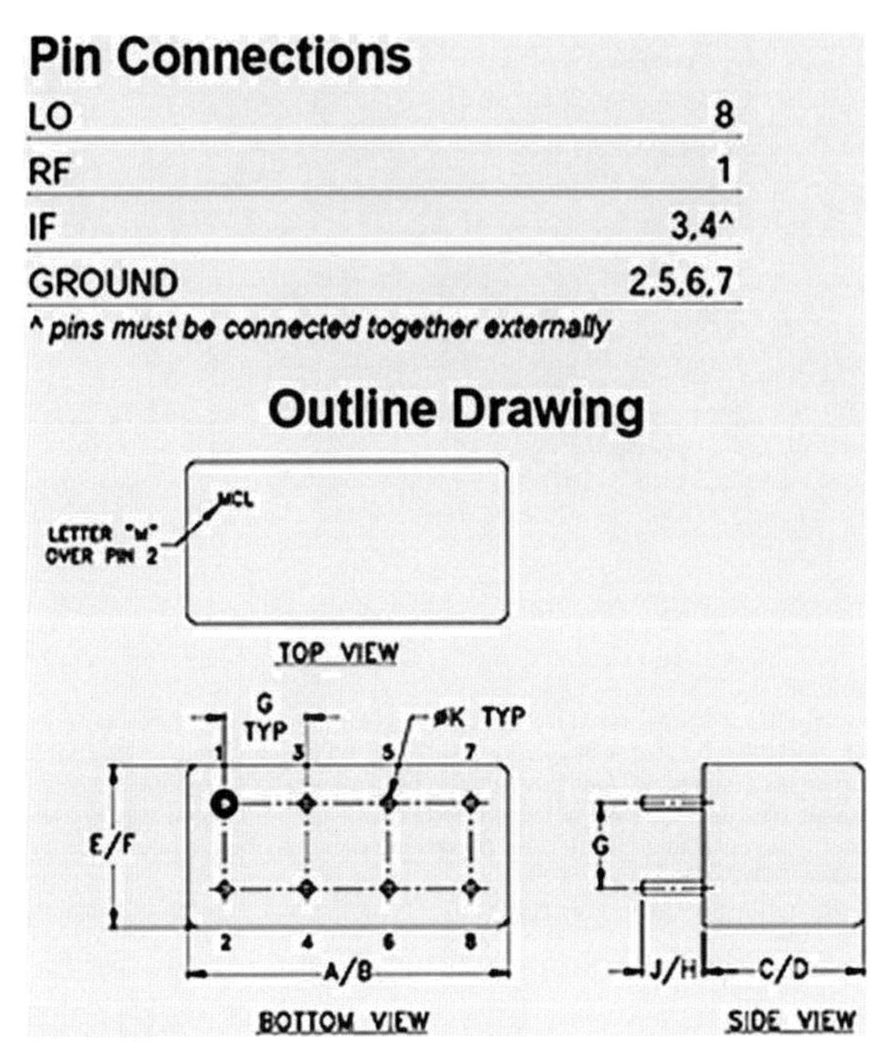

Pin Connections

LO	8
RF	1
IF	3,4^
GROUND	2,5,6,7

^ pins must be connected together externally

Fig. 12.5. Pin configuration of Minicircuits SBL-1+.

The IF Filter

The down converter circuit illustrated here outputs 10.7 MHz. This is a common IF frequency used in television and commercial radio receivers, so there are many filters available readymade to clean up the output of the mixer before it is processed by the next stage. These come in ceramic or crystal forms, but are simple – a package with usually three pins, although some crystal types come with four pins because they have two ground pins. Murata makes a range of them with varying bandwidths of a few kHz to over 400 kHz. In practice use these wherever you can. However for more general filtering needs where filters are required to operate at custom frequencies or non-standard bandwidths you can build your own using groups of inductors, capacitors, and sometimes resistors. This is covered later.

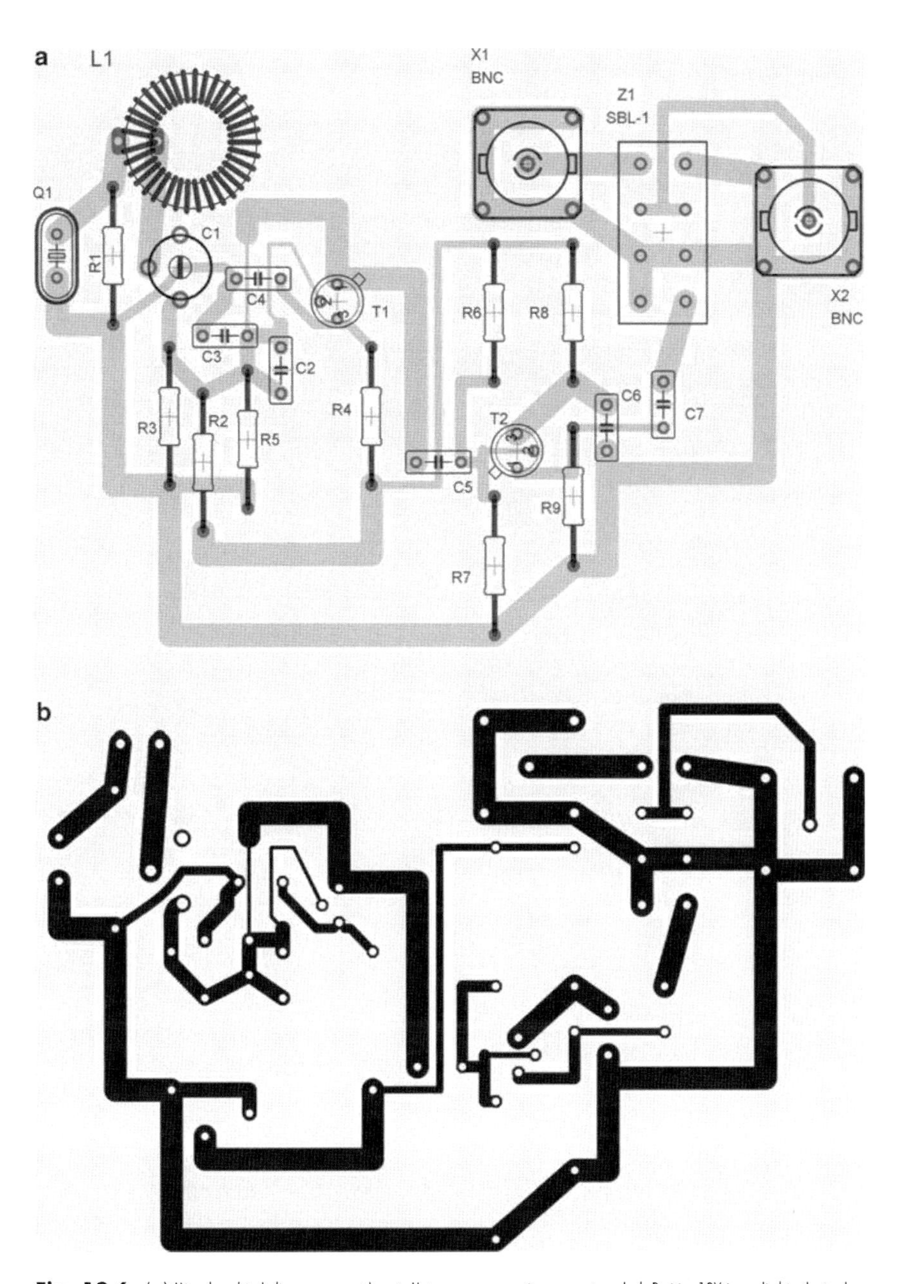

Fig. 12.6. (**a**) Mixer board including component layout. Note: power connections are not marked. Positive 12V is applied to the track connecting R2,R4,R6 and R8. Ground is the track connecting R1,R3,R5,R7, and R9. Confirm this by comparing with Fig. 12.3 (**b**) Mixer board ready to print onto heat transfer. The layout suggests the BNC sockets are attached to the board. In practice use chassis mount sockets and connect to the board with a short length of wire.

In the prototype for this project a Murata 3 pin ceramic 10.7 MHz filter with a 330 kHz bandwidth was used. Since this mixer is passive, there is a loss of around 7 dB. So it is usual to add an amplifier stage after the mixer. The amplifier used was based on a MAR6 chip, also available from Minicircuits. The circuit is shown in

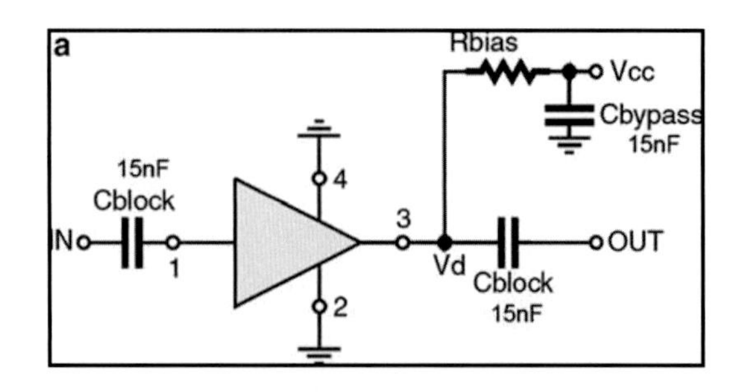

Fig. 12.7. (**a, b**) An RF amplifier used at the output of the mixer, constructed on a small piece of strip board. The pair of resistors made 660 Ω for a 14 V supply. The signal is input via a 3 pin ceramic filter (*lower center*). Note that a piece of copper was removed from beneath the Mar6 chip.

Fig. 12.7. The input pin 1 is marked with a small dot. It is a very small surface mount component, but can be soldered onto the board with a fine-tipped soldering iron while holding the chip down with a miniature screwdriver. The prototype was constructed on a small square of strip board. A small circle of copper was first removed from around one hole, and the chip was soldered such that an earth strip bordered both sides, and the input output pins straddled the broken strip.

The bias resistor is chosen to suit the power supply; the values can be found in the data sheet obtainable from the Minicircuits website. The gain is around 22 dB, which is more than enough to overcome the loss in the mixer stage.

The Logarithmic Amplifier

The final stage is the logarithmic amplifier. It is based on a series of amplifier modules, but its purpose is not to amplify a signal but to be a measurement device. The output of a log amp is a DC voltage whose value is directly proportional to the power of the RF signal applied to it. The output is scaled logarithmically. This offers a way of compressing a signal with a wide dynamic range into a compact logarithmic scale, which is easier to view. Since I'm recommending the Analog Devices AD8307 chip we will not go into its internal function in any detail. If you want to know more about it download the data sheet from the Analog Devices website. You should treat the chip as a "black box," a device that you can apply in practice but don't necessarily understand what goes on inside.

The AD8307 is a very simple device with only eight pins, and you can get it in the through-hole style; a suffix of N on an IC chip normally means through hole pin style. The package is also known as PDIP, for plastic dual inline package. The Table 12.6 details the pin functions. IC pin number 1 on any PDIP IC can be identified by either a small circle marked beside it, or by a semicircular indent in one end, where if the indent is held uppermost, pin 1 is on the top left. The numbering is always down the left side and continues in a vertical manner up the right side, so pin 8 is opposite to pin 1 in this case (Table 12.6).

The AD8307 requires very few external components to function. It has a dynamic range of 92 dB and will operate with signal power levels from -78 to +17 dB. The output voltage will vary from 0.25 to 2.5 V, so that 25 mV represents 1 dB change in signal power, assuming a power supply of 5 V. This would drop to 15 mV/dB at 2.7 V. Since the tuner requires 5 V, it makes sense to supply the log amp from the same power supply. Therefore an 8-bit data logger is perfectly adequate (see Chap. 13).

The Log Amp Detector Circuit

The circuit diagram is shown in Fig. 12.8.

The PCB layout for the log amp is shown in Fig. 12.9. Pad 1 output and ground can then be fed to an analog-to-digital converter for logging the data, or to a digital voltmeter for system testing.

If this module is built in a separate box of its own, it can be used to interface other instruments you make. It will work from audio frequencies up to 100 MHz to full specification and will continue to operate up to 500 MHz with slight reduction in performance. So this could be used on its own connected to the audio output of a conventional radio receiver as a way of logging changes in signal power.

Antenna Choice

If the tuner front end was restricted to the UHF band of operation, then the easiest solution would be to buy a high gain television antenna. If the full range of frequencies were required then a log periodic array would be more suitable, at the expense of reduced antenna gain.

Table 12.6. The AD8307 log amp pin connections

Pin number	Mnemonic	Description
1	INM	Negative signal input
2	COM	Common pin usually grounded
3	OFS	Offset adjustment
4	OUT	Logarithmic output voltage 25 mV/dB
5	INT	Intercept adjustment
6	ENB	Chip enable, turn high to activate it (apply power supply voltage)
7	VPS	Power supply 2.7–5.5 V
8	INP	Positive signal input

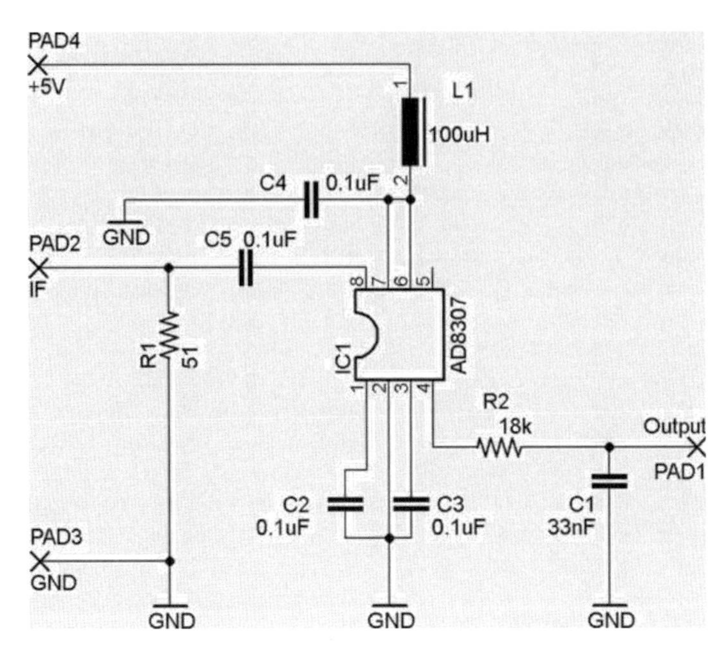

Fig. 12.8. Logarithmic amplifier/detector.

DIY Filter Designing

To design a filter from scratch involves a fair bit of tedious mathematics. Not that the mathematics is very difficult. It's just that there is a lot of room to make a mistake. We will not subject you to an early evening's sleep by trying to explain the mathematics here, but there is a rather good explanation of one method of filter design in the book *RF Circuit Design* by Chris Bowick. What is recommended is to download a cool piece of freeware software simply called Filter Design. You can get it from AADE via the website http://www.aade.com. There are lots of free filter design tools and some decidedly not so free ones, but this is one of the best. When you start the software there is a small advertisement for AADE's own LC meter, which is an invaluable tool in its own right.

Once you open the program you get a window with a help box displayed. Start by selecting the design menu, and then Butterworth. We will not go over all of the features of this software – you can of course explore it more fully yourself. But let's cover an example design the way you might intend to use it.

The terms Butterworth, Chebyshev, etc., refer to design styles; each has its own particular characteristics. Butterworth types tend to have a very flat pass band response, while Chebyshev types can show a fair amount of ripple. By selecting Butterworth you are asked what kind of filter you wish to design, bandpass, low pass etc. Select bandpass, so we can design a filter for the output of our mixer, and press OK. You are now presented with a dialog box, as in Fig. 12.10.

The box CF is where you define the center frequency, in this case 10.7 MHz. BW is for bandwidth. Now, making very narrow band filters is not a good idea using Butterworth principles, but that's OK. Let's chose 500 kHz. Next set SF, or shape

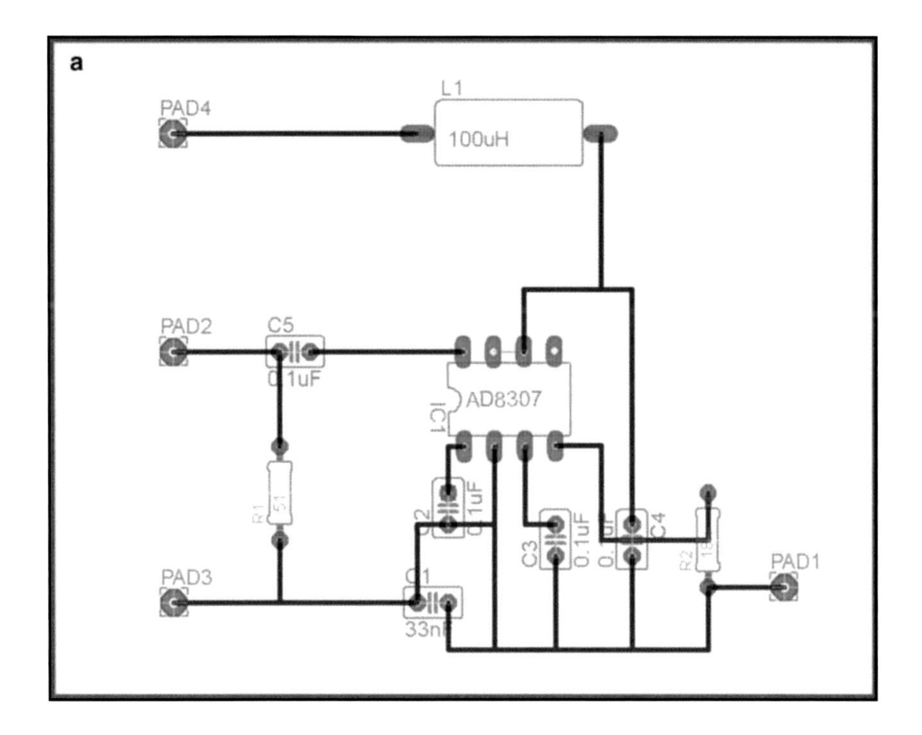

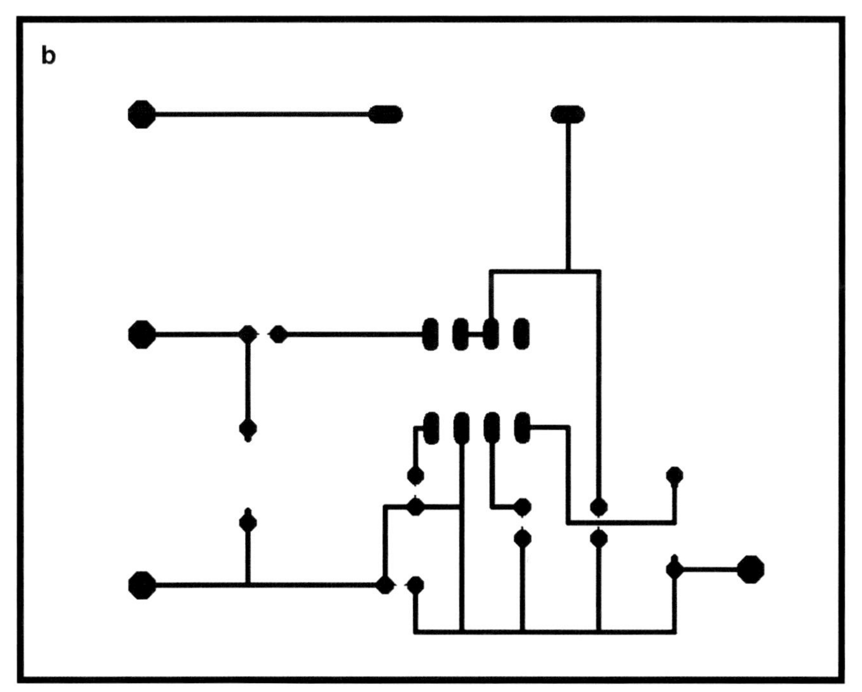

Fig. 12.9. (**a**, **b**) Logarithmic amplifier PCB layout.

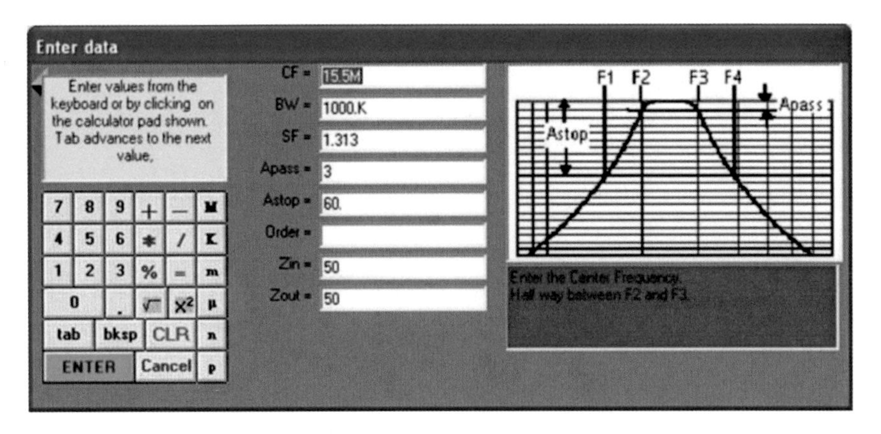

Fig. 12.10. Dialog box from Butterworth.

factor. When you click in the SF box help appears in red under the graph that explains what it is. SF defines how fast the roll off is in the design.

Very steep-sided designs are unreasonable, but the F1 and F4 frequencies are chosen to define the attenuation required at a specific distance away from center frequency. Now there is an amateur radio frequency band operating at up to 10.15 MHz, so let's make sure we attenuate that strongly. The F2 is now 10.7–0.25 MHz, which is 10.45 MHz, and F1 is 10.15 MHz, so SF is F2/F1 = 1.03. APass is chosen to the point where the power is reduced by 3 dB, so enter 3 there. AStop is how much attenuation we want at frequency F1, so let's make that 30 dB as a first try anyway, so enter 30 in the box. Leave order blank; the software will decide this value. Order effectively means how many components are required. Finally enter the input and output impedances in ohms. These can be different, but keep them to 50 for this exercise. (As an aside, note that a filter can also be used to match the impedances of dissimilar stages.) Now press enter. And enter again for a lossless design. You get the chance to name your design, and then the filter is shown as a schematic with component values. That's it – job done! Hmmm, maybe not. For one thing the values of the components are unlikely to exactly match real-world values.

Next step is to analyze the design, so click on the Analyze menu, and select the top item power gain. In the resulting dialog box set the upper and lower frequencies for the analysis. In this case say 9 and 12 MHz, and set the attenuation range from 0 to 50 and click enter. The resulting graph should look like Fig. 12.11.

The graph represents the attenuation of RF power over a range of frequencies. Unwanted mixer products will be removed from the output, and a good degree of attenuation is applied to potential sources of nearby interference.

Now examine the component values calculated. Figure 12.12 shows the schematic and the values.

Since you can't make your capacitors but you can make your own inductors, you could try changing the capacitor values to those that you can buy and re-analyzing the design to see what effect it has. Then adjust its neighboring inductor value to bring the power gain plot back to where it was. Do this one component at a time, using only small changes all through the circuit. The resulting inductors can be constructed using toroid cores in the manner explained in Chap. 8. However some designs can be very sensitive to tiny tolerance changes in values. It is not a good

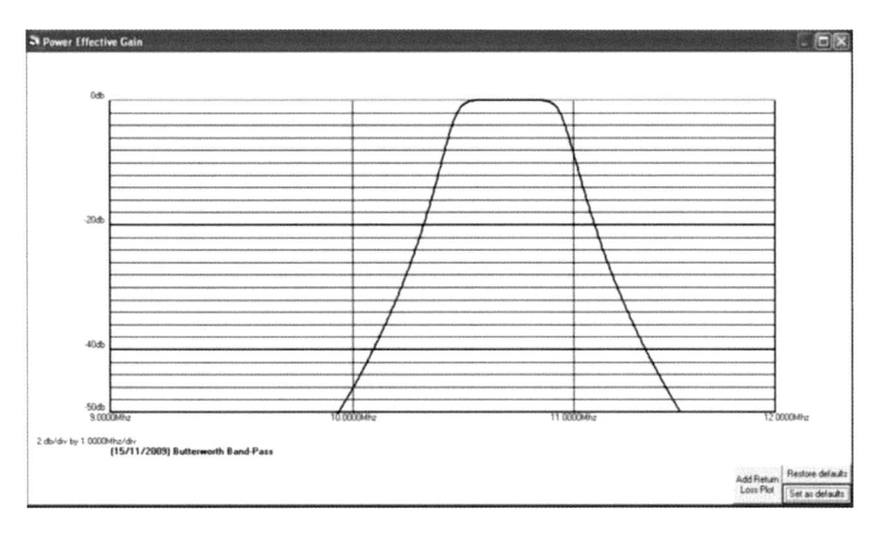

Fig. 12.11. Attenuation of RF power over a range of frequencies.

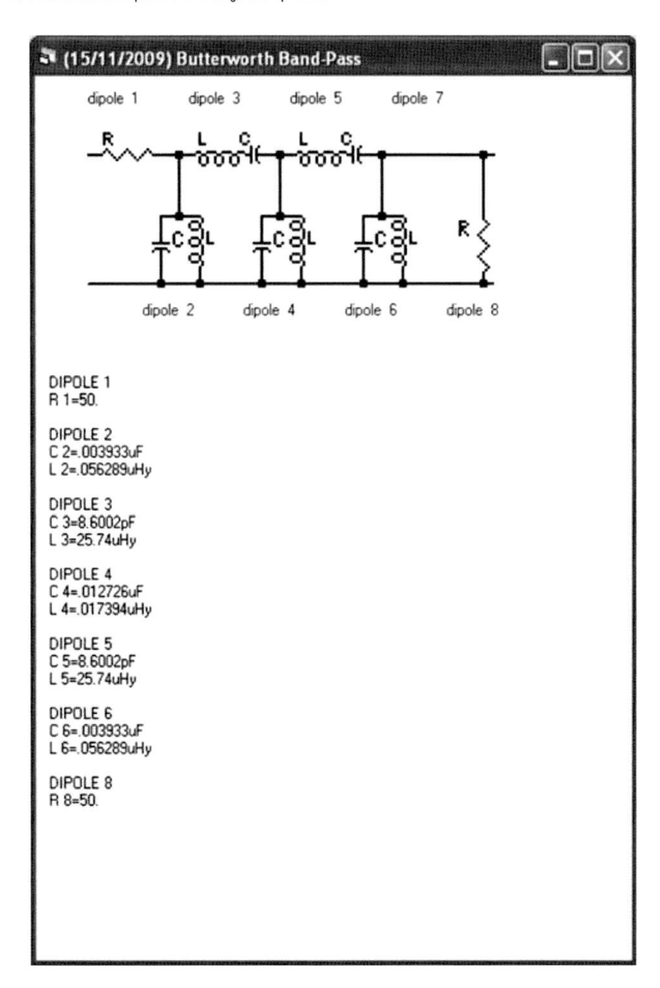

Fig. 12.12. Schematic of component values.

idea to design these simple filters, which have a very narrow bandwidth and a sharp cutoff. Be realistic about bandwidth and roll off. Experiment with the software until you find a design that is practical.

When building the filter, use screening plates between each dipole of the circuit. If the filter is constructed "ugly" style onto double-sided copper board, then pieces of double-sided copper board can be soldered to the base, mounted vertically to act as screens.

Use this technique for preselectors, mixers, IF amplifiers, and even audio circuits. These can all be analyzed for practicality before any components are even purchased.

Converting to a Spectrometer

There are at least three basic ways to build a radio spectrometer: the scanning receiver, the filter bank receiver, and the multichannel receiver.

The scanning receiver is the simplest. In the case of this project software can be written on the PC that causes the tuner to scan through its range. A single scan will provide a radio spectrum of the range of frequencies used in the scan. This project is very much an experimental bench test. PC software discussion is beyond the scope of this book. It is left to the reader to explore ways of developing software for this purpose. If you refer to Chap. 13 where the Arduino hardware is discussed, not only is this a suitable solution for data logging the output of the log amp but the PC software Processing should make it easy to code applications capable of tuning and receiver and to make it scan frequency ranges. To help learn the Processing language the book called *Processing: A Programming Handbook for Visual Designers and Artists* by Casey Reas and Ben Fry is highly recommended.

The filter bank spectrometer simply sends a copy of the signal to a bank of filters, so the output from each filter is a small part of the whole spectrum, and the group of filter outputs make up the complete spectrum. The advantage of this technique is that a spectrum is built up instantaneously, where a scanning receiver takes time to build up the complete spectrum. This is not suitable for this receiver due to its restricted bandwidth.

The final type is the multichannel receiver. A group of identical receivers could be set up on different channels to instantaneously record signals, avoiding the time-delay problem of a scanning receiver. It would not be a big advantage to have large numbers of receivers on contiguous channels, but it would be useful to have 6–8 set up to cover channels evenly spread across their operating range. Since many solar flare events involve bursts of radio energy several megahertz in bandwidth drifting in frequency, the passage of the burst could be followed in the data from the separate channels. Many of the recommended data logging devices have 6–8 channels available, so only one logging device is needed in this project. The total cost of one of these receivers is fairly small, so it is a viable way of studying the time and frequency dependence of solar flare events.

Chapter 13

Data Logging and Data Processing

After constructing a radio telescope consideration needs to be given on how to log and process the information your receiver is gathering. The old way used paper chart recorders, which have been made obsolete now by the personal computer. Even an old obsolete computer can be valuable tool for the electronics bench or as a processing and control unit for a radio telescope. Old unused PCs can often be obtained free of charge and put to use in a second life. PCs can be energy hungry, so there is an intermediate step that is possible, the stand-alone data logger. A low-power device based around a microcontroller and some flash memory used to constantly record the data from the telescope. Data can be dumped once a day or so to a PC for further processing.

Here we will look at easy ways to log data. Microcontrollers are a valuable tool for instrument control and data logging. Constructing your own custom devices is certainly within your capability and is often much easier to achieve than building analog RF electronics. Space does not allow a full enough discussion on the subject, so it is recommended that you read up on the subject later. However, readymade data logging tools are available to buy. Here are some options.

The Logging Multimeter

The photograph shows a Tenma model 72-7732. Although it is not the cheapest model in its category, it has some useful features (Fig. 13.1).

For the purposes of data logging, it comes complete with a USB interface and data logging software. This relies on the personal computer to log and store data. Since a radio telescope back end usually provides a DC output signal an instrument like this acts as an analogue to digital converter, and a PC interface.

In addition to this as well as the usual DC, AC voltage and current ranges, diode tests, etc., it can measure capacitor values and has a useful frequency counter operating at up to 400 MHz. This is useful for setting up oscillators in radio circuits. The only down side to having only one of them is when you come to build the next radio telescope! To use it as a test instrument you have to stop logging data.

There are lower cost models around for as little as one fifth the price of the Tenma, which typically have an RS232 serial interface in place of the USB. This

J. Lashley, *The Radio Sky and How to Observe It*, Astronomers' Observing Guides,
DOI 10.1007/978-1-4419-0883-4_13,

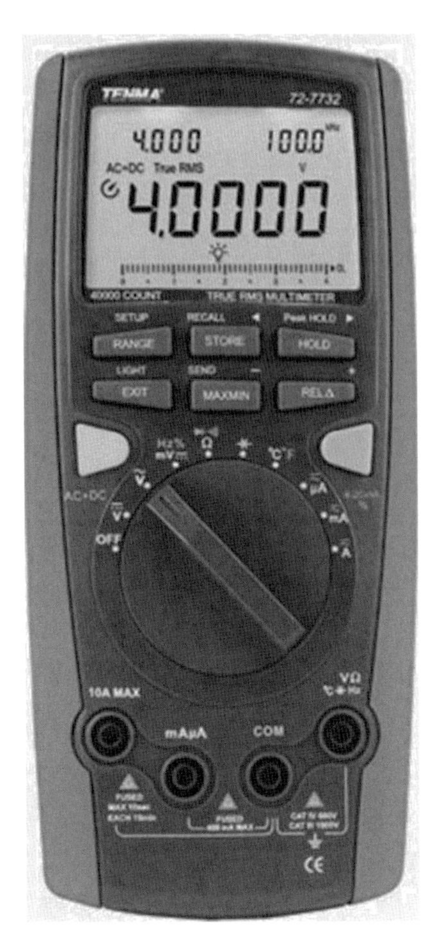

Fig. 13.1. Tenma model 72-7732.

would suit older PCs or may work via a USB to serial interface. One model looked at was an "unbranded" unit sold via eBay. It also had a frequency counter but would only work at up to 10 MHz, much less useful. It is always worth spending a little more money, or a lot more money if you are serious, to get a better quality instrument that you can rely on for accuracy and robustness.

More Data Loggers for the PC

Pico Technology provides a wide range of data loggers and PC-based instrumentation. Shown here is the high-resolution 24-bit data logger. For more details see their website, http://www.picotech.com (Fig. 13.2).

The limitation of this model is it can only handle a range of ±2.5 V. The output of the radio telescope must therefore be scaled to operate within this range.

Fig. 13.2. High-resolution 24-bit data logger from Pico.

All Pico data loggers come with a software application called PicoLog, which handles the data capture to the PC and provides visualization in the form of graphs.

These units work out about twice the price of the Tenma meter, so maybe you should buy two Tenmas? Well, maybe not. The Pico tool will certainly provide a much better resolution for capturing fine detailed changes.

The other advantage of a tool like the ADC24 is the ability to write your own software applications and embed the functions of the unit into your program. The manual supplied gives all the information you need.

The Home Brew Method

For those of you who really want to feel you have built the system yourself, there are simple ways of constructing your own logging interface. Together with some low cost software available via the Internet you can have an eight-channel logging solution. More than one channel is useful if you have multiple instruments to log, say, by monitoring more than one radio frequency simultaneously. Since the performance of radio receivers can be influenced by ambient temperature, you may wish to build temperature sensors into your instruments and log the temperature alongside the data. Later calibration adjustments can be applied based on how the temperature varies.

A useful microchip is available from Maxim, the MAX186. Available in DIP format (dual inline package), it has 20 pins. Its data logging functions work without the need to write and store software onboard the chip, so it is an off-the-shelf solution. The chip can be configured a number of ways to talk to a computer, but the application here shows how to connect it via an old style printer port to a PC (Fig. 13.3).

Note the convention used in the circuit diagram shows a dot where lines connect. If lines cross each other without a dot being present, they are electrically

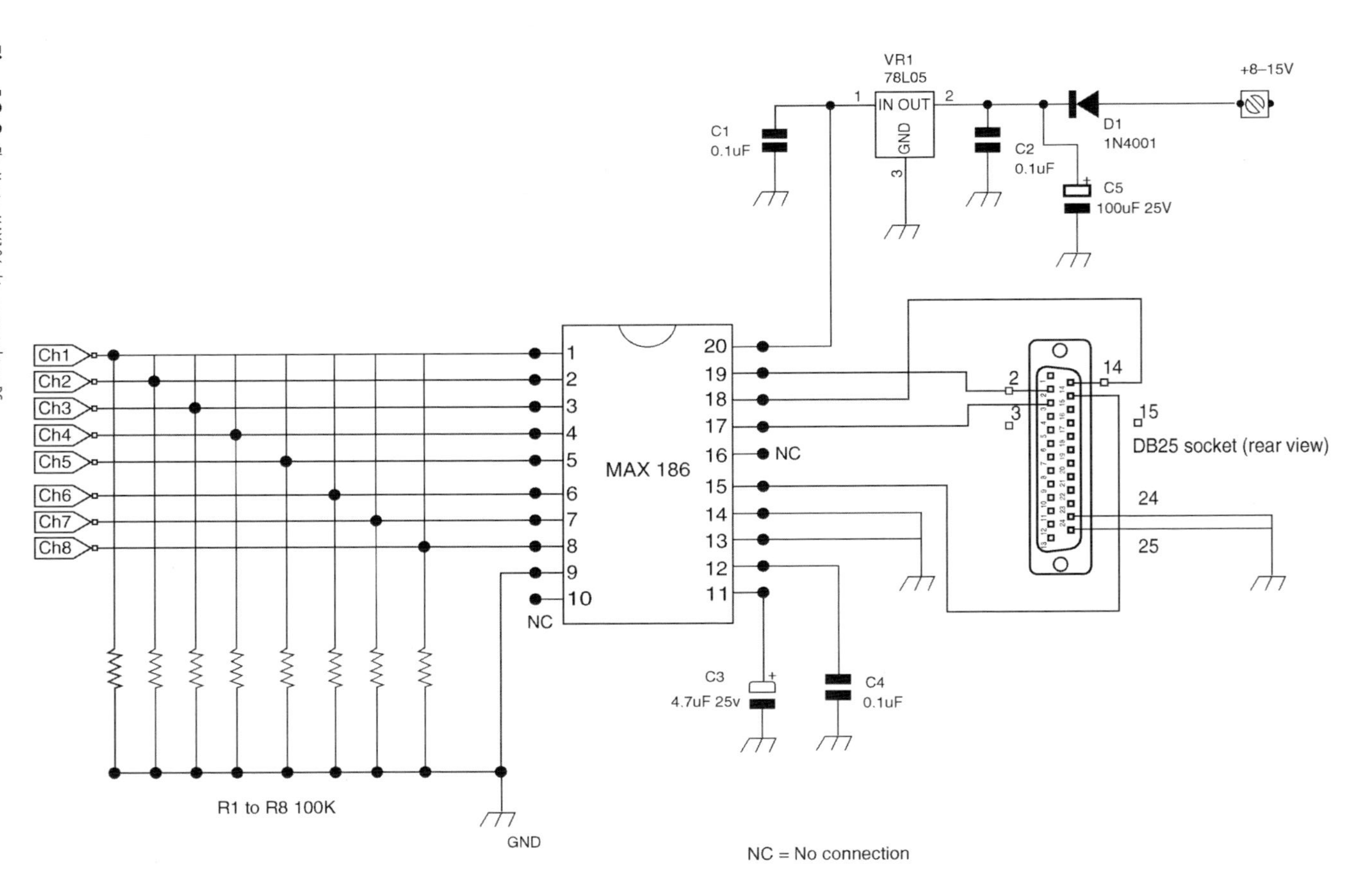

Fig. 13.3. The Maxim MAX186 chip connected to a PC.

isolated from each other. The circuit can be constructed onto strip board but the following figures show the PCB layout (Fig. 13.4).

Radio *Skypipe* Software

The previous logger design is particularly suited to the radio *Skypipe* software available for evaluation from http://www.radiosky.com. The evaluation version has a few limitations but is fully functional as a single-channel logging option. The full version is relatively inexpensive and allows the full eight channels to be logged; it also opens up the possibility of streaming data via a live website.

Skypipe is simple to use and is the modern equivalent of a chart recorder, storing its data instead to a computer drive. This is another highly recommended piece of software for the beginner.

Figure 13.5 shows a VLF receiver plot taken on June 8, 2007, showing a solar SID event.

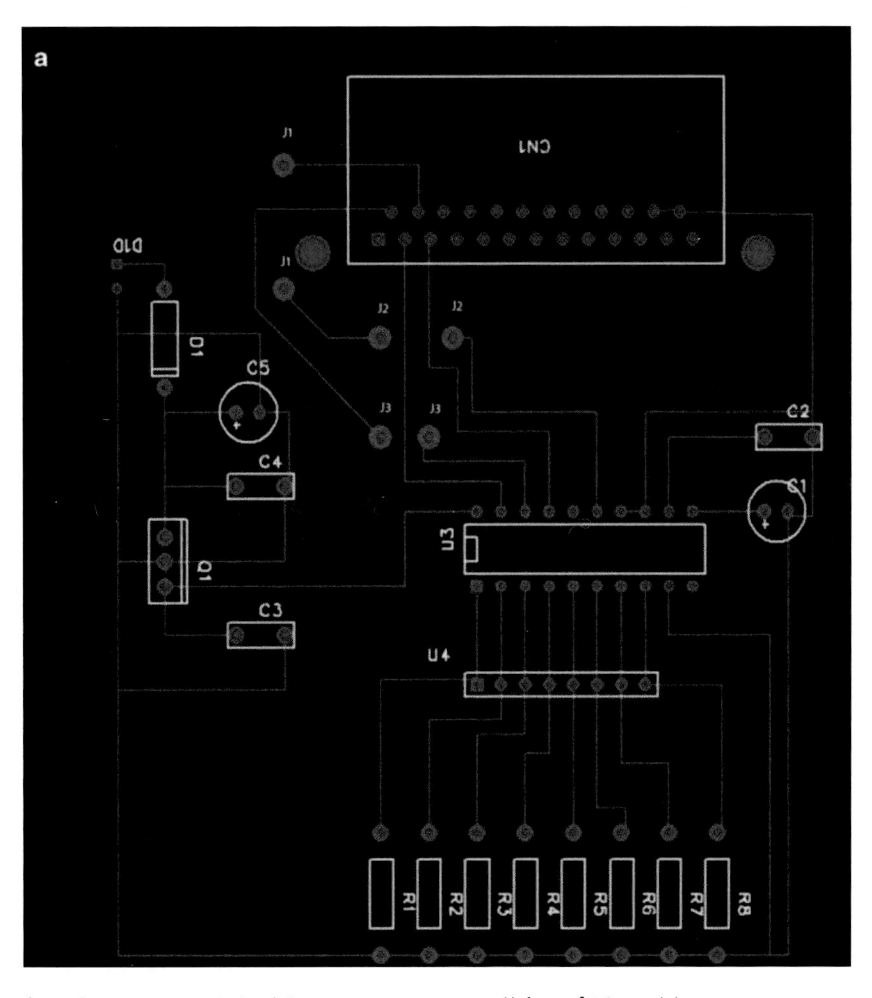

Fig. 13.4. (**a**) Top view: J1, J2, and J3 connection points require jumper cable fittings. (**b**) Copper side bottom view.

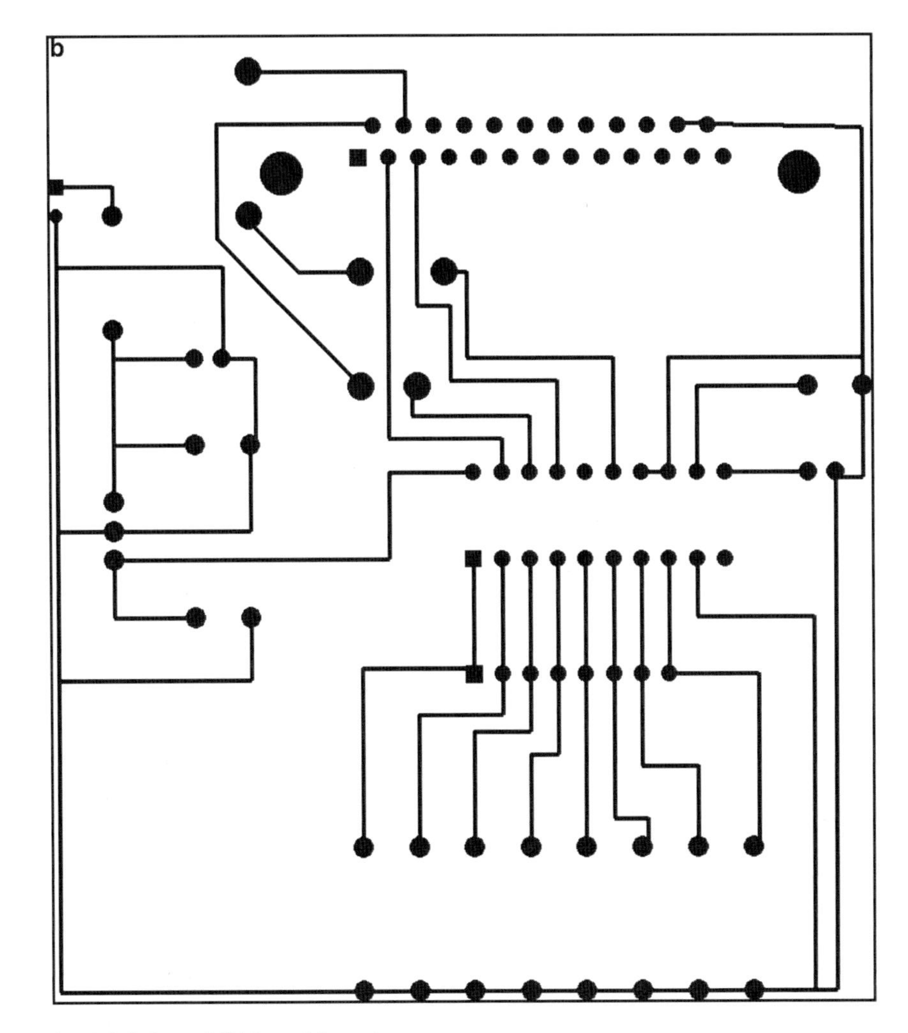

Fig. 13.4. (continued) (**b**) Copper side bottom view.

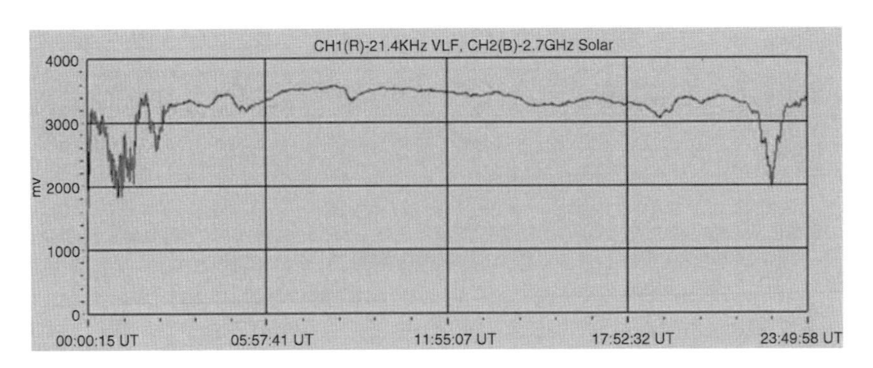

Fig. 13.5. Solar SID event logged using Radio *Skypipe* software and a VLF hardware receiver. The image is dated June 8, 2007. It shows a typical SID disturbance at around 08:00. (Image courtesy of Martyn Kinder from the British Astronomical Association Radio Astronomy Group).

The Controller Option

The controller is a more sophisticated option than we have seen so far. By controller, we mean a small single-board low-power computer based on a microcontroller. There are many available ready-built. The advantage of a controller is that it can be used to control the operation of the radio telescope, as well as handle data capture.

The controller used by the British Astronomical Association Radio Astronomy Group is the Futurlec ADuC832, available from Futurlec direct via their website, http://www.futurlec.com. This has an on-board microcontroller manufactured by Analog Devices, the ADuC832. The down side here is that you will need to create your own embedded software for it. Example software is provided with the kit (Fig. 13.6).

The features on this controller include 64K of flash program memory. (This may not sound like much, but it's surprising what can be accomplished with that amount of space.)

The software can be downloaded from a PC using a serial lead. It has eight channels of analog-to-digital conversion and two channels of digital-to-analog conversion, plus up to 1 Mb of RAM can be installed as an option. It has a real-time clock on board and can be made to communicate over very long distances using RS422 or RS485 by the addition of low-cost chips.

It would be possible to control the Philips CD1316L of the solar radio telescope project tuner via I^2C using this controller, as well, of course, to log the output data. With the proper software configuration the Philips tuner could easily be made to sweep quickly across a range of frequencies to turn the instrument into a spectrometer.

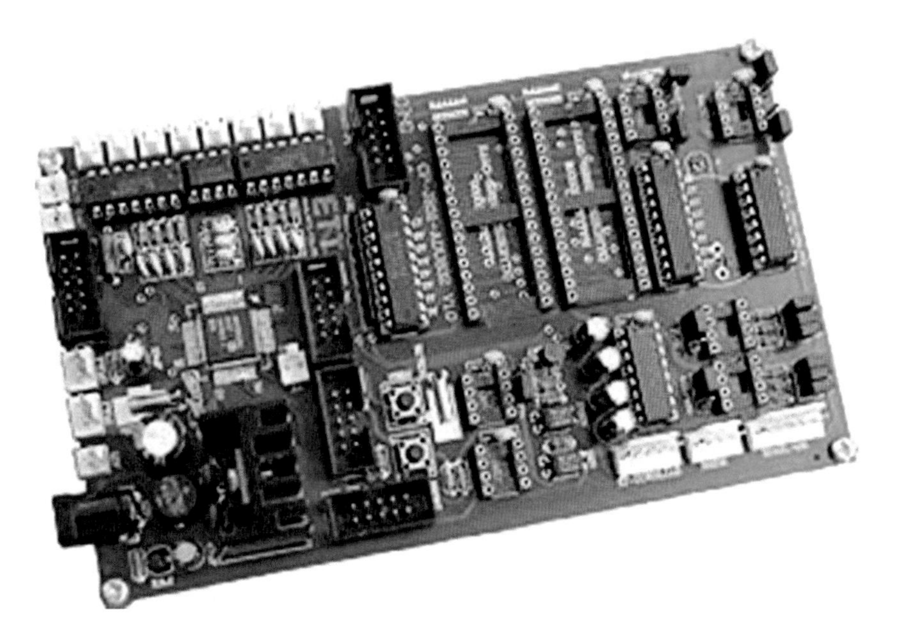

Fig. 13.6. The controller from Futurlec.

The Arduino Duemilanove

Take a look at the Arduino; it is one of the most engaging and addictive bits of technology around. It comes in a number of variations, and there are some clones such as the Seeeduino (yes, three e's). There is nothing particularly special about the hardware itself. What makes the Arduino a highly recommended platform is the community behind it. The software behind the Arduino concept is all open source and free. There are many individuals and groups using it for all kinds of projects, so there are lots of coding examples to choose from and adapt to your own projects. This is always a great start when something is new to you (Fig. 13.7).

Developing software is beyond the scope this book, but this is such a useful piece of hardware that included here is some sample software to get you going. This is truly one of the easiest microcontroller platforms to work with. There are also lots of ideas available to help interface this hardware into well-known software packages such as Flash, Visual Basic, Matlab, Java, and many more. The development environment is itself based on an open source programming language called Processing. Processing was developed to be easy to learn and use for teaching programming and as a platform for designers (rather than programmers) to be able to create visual software. Processing is an ideal language to learn if you are not already a programmer, to develop your own applications to work with the Arduino hardware.

The Duemilanove (meaning "2009" in Italian) has the following features:

- 14 digital input–output pins
- 6 of the above pins can be made to output PWM (useful for controlling motors)

Fig. 13.7. The Arduino Duemilanove controller.

- 6 analog input ports, 10-bit resolution (giving 0–1,023 for up to 5 V inputs)
- 32K of Flash RAM for programs using the ATmega328
- 2K of SRAM
- 1K of EEPROM

When you compare the memory capacity of a microcontroller with that of a PC it seems tiny, and you wonder how this could possibly be useful for anything. Well, it goes a long way. If the primary aim is to provide an analog-to-digital interface for a PC, then an average program running on the microcontroller will probably take up about one-third of the available space or less. Most of the sophistication is handled by the PC.

Communication with the PC is by USB cable. Drivers are supplied that create a virtual COM port, making serial communications easy between the two. In fact the FTDI chip is already known to Microsoft Windows and should not pose any problems to install.

Another advantage of using the Arduino is the ability to piggyback hardware on top. Piggyback boards are known as shields. There are shields that can add SD card slots for huge storage solutions, Real-time clocks, even wireless or wired LAN boards, are available.

To get started download the Arduino programming software from the home page at http://arduino.cc and uncompress the archive to a folder on your computer. Then download the Processing software from http://processing.org and uncompress it to a folder of its own. The two environments look alike, because the Arduino software is based on Processing. It is easy to confuse the two when you are developing PC/Arduino applications!

To set up the Arduino software you need to tell it which serial port to use. The FTDI chip emulates a serial COM port on the PC. It may appear as COM3 or COM5 but may be even higher if you have used other similar devices before. In the menu "Tools/Serial Port" click over the port used for the Arduino. If you are using a modern PC or laptop it may be the only COM port available anyway. Under the menu "Tools/Board" you can set the type of Arduino you are using.

Next type in the following program to upload to the microcontroller:

```
// Code to read the analog value supplied from analog pin 0
int val;
int inputPin=0;      // Set analog input pin 0

void setup(){
 Serial.begin(9600); // Star serial communication at 9600 bps
}
void loop(){
 val = analogRead(inputPin); // Read the input pin range 0 to 1023 10 bit
 Serial.print(val/4, BYTE); // Convert to 8 bit and send via serial port
 delay(100);           // Wait 100ms
}
```

At the top of the window is a row of buttons. Click on the one square button with the right arrow in it, the upload button. This first compiles the software then programs the chip on the Arduino. You will see leads flash on the board when it is programming. Once they stop a text message appears on the PC screen. If you make errors in typing you may get warnings on the screen. Correct them, checking against the above code, and try again. When completed press the reset switch on the Arduino board, and close the PC software.

Next open the Processing software and type in this program:

```
/**
 * Simple Analog Data logging application using Arduino hardware
 * Samples are taken every second.
 * Data is read from the serial port, and displayed live as a bargraph
 * while at the same time is saved as a comma separated text file
 */

import processing.serial.*;          //import serial library

Serial myPort;                        // Create object from Serial class
int val;                              // Data received from the serial port
String d = str(day());                //Set current date
String m = str(month());
String y = str(year());
PFont font;                           // create a font object
String[] dataout = {"logger data "+d+":"+m+":"+y}; //Create String array with header

void setup()
{
 size(300, 600);                     //Define window size
 font = createFont("Arial",22);
 textAlign(CENTER);
 textFont(font);
 println(Serial.list());             //Prints a list of available COM ports
 String portName = Serial.list()[0];     //Change the [0] value for your COM port
 myPort = new Serial(this, portName, 9600); //Open serial port
}

void draw()
{
 if ( myPort.available() > 0) {          // If data is available,
  val = myPort.read();                   // read it and store it in val
  delay(1000);                           // delay 1 second
 }
 background(255);                        // Set background to white
  fill(0);                               // set fill to black
  rect(30, height, 240, -val*2);         // create bar graph rectangle
  text(val,width/2,height-val*2-20);     // Add text value above

  // create string to add to array
  String s = str(hour())+":"+str(minute())+":"+str(second())+","+str(val);
  dataout = append(dataout, s);          // append string to the array
  saveStrings("data.txt", dataout);      // save the array to a text file
}
```

Note that the // symbol is used to add comments to help you understand how the software works. When ready save the software and click the Run button at the top left of the Processing software. A window is then created, displaying a real-time bar graph of the data being streamed from the Arduino. At the same time the data is saved to a text file called "data.txt" that appears in the same folder in which you saved the program. The current date is appended to the header text, and the current (PC) time is appended to the data value. This text file could be imported into Microsoft Excel or similar for data processing and graphing, or you could develop a more sophisticated Processing application to do the same thing.

This software is intentionally simplistic, so hopefully it is easy to understand. It needs further work to be a practical tool, but it gets you started. Note that the data is captured as a 10-bit value (0–1,023) but is compressed into an 8-bit BYTE (by dividing by 4) before sending from the Arduino. The 10-bit value takes up four BYTEs, but it also makes the sensitivity very high. For example the sensitivity for a 10-bit ADC sample of a 5 V signal means 1 bit is about 5 mV. At that level even the 5 V power supply from the USB port showed extensive ripple on the order of 15–20 mV. By converting to 8-bit the output is much smoother, and 1 bit represents 20 mV. You need only code a single channel, but there are six analog input ports; so you could modify it to capture all the channels you need. It is up to you to experiment with it.

The PC Sound Card as a Data Capture Device

We have seen an example of using the PC sound card as a data capture device. The line input port of a sound card is simply another form of analog-to-digital interface, and so is the microphone port. Although the microphone port is capable of measuring very small signal levels, due to the existence of onboard preamplification within the sound card, it is the line input port that will be most useful.

The definition of line input level varies between consumer and professional equipment. Consumer equipment is usually set up to handle up to 1 V peak to peak, and professional equipment about 1.7 V peak to peak. Assuming the sound card is intended for the consumer market that means the peak amplitude the card will deal with is only 0.5 V. A circuit could be constructed to scale a 5 V signal down to 0.5 V, but it is counterproductive to do so. Where a sound card comes into its own is when a conventional radio is employed with an audio output. There are several software applications out there that can process an audio signal in order to extract useful information. The example recommended is Spectrum Lab released by Wolfgang Buescher, whose amateur radio call sign is DL4YHF. The website where you can download it is http://freenet-homepage.de/dl4yhf/spectra1.html.

One of the main uses of Spectrum Lab is a processing technique known as FFT, or Fast Fourier Transform analysis. A generic sound signal provides us with a variable waveform whose amplitude varies with time. At any given instant, the amplitude of that signal is a sum of the all the amplitudes of all the different tones or frequencies that make up a sound. Human hearing can in theory detect sounds with frequencies up to 20 kHz. In practice for most people it will be less than this.

Our hearing changes with age, and the upper limit of frequency detection falls. This is of course not important, as the PC is going to do the analysis for us. What it means in practice is that a PC sound card has a flat response up to 20 kHz, often 21 kHz, and then its response rolls off steadily and falls below ambient noise levels between 23 and 24 kHz for most cards (assuming a 48k/b sampling rate).

To picture how a sound card is useful for us, consider how the AM, or amplitude modulation, radio works for a moment. The AM radio signal is an amplitude varying RF signal, and the information it carries depends on how the amplitude varies. The amplitude "envelope" of an AM communications signal is in fact the audio signal superimposed onto the RF wave. Figure 13.8 illustrates this.

The radio translates this varying amplitude to a low frequency. The translation process of the radio removes the radio frequency "carrier" and provides a time-varying audio signal equivalent to the profile of the original wave. The FFT spectrum generated by Spectrum Lab converts this time-varying signal into the frequency domain. The plot then displays how amplitude varies with frequency. Any arbitrary waveform can be represented by a series of pure sinusoidal components of different frequencies and amplitudes. The branch of mathematics involved with converting time-varying signals into the frequency domain is Fast Fourier Transform.

Therefore a sound card can turn an ordinary radio into a narrowband spectrometer, and by tuning the radio the spectrum can be scanned across a wide range. This is not only potentially powerful but very low cost. The chances are high, if you are reading this book that you already have the PC and sound card. What's even better is that Wolfgang offers his software free for personal use.

The basic principles of using Spectrum Lab were covered in the VLF receiver chapter. But Spectrum Lab is capable of much more. It has a built-in scripting language that you can use to automatically capture spectra of plots when your test conditions are met. For example if this was used to count meteors, the script could

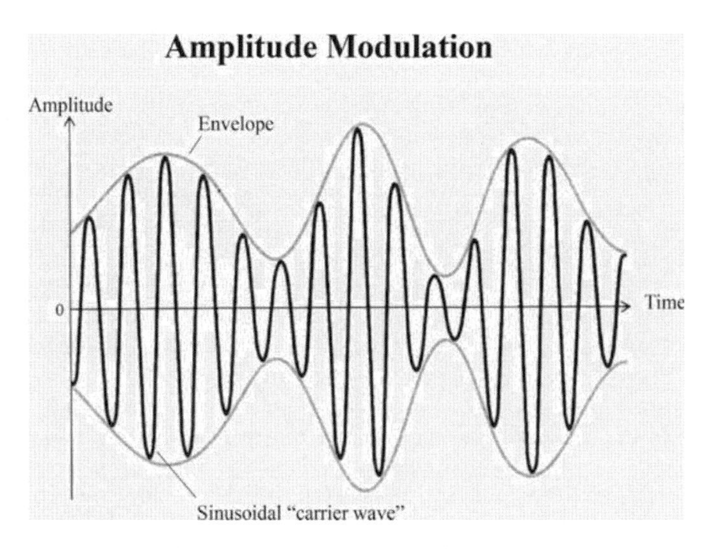

Fig. 13.8. Amplitude modulation of a sine wave. In this example the envelope of the wave is a single frequency audio tone superimposed on a higher frequency carrier. In practice the carrier is much higher frequency than shown.

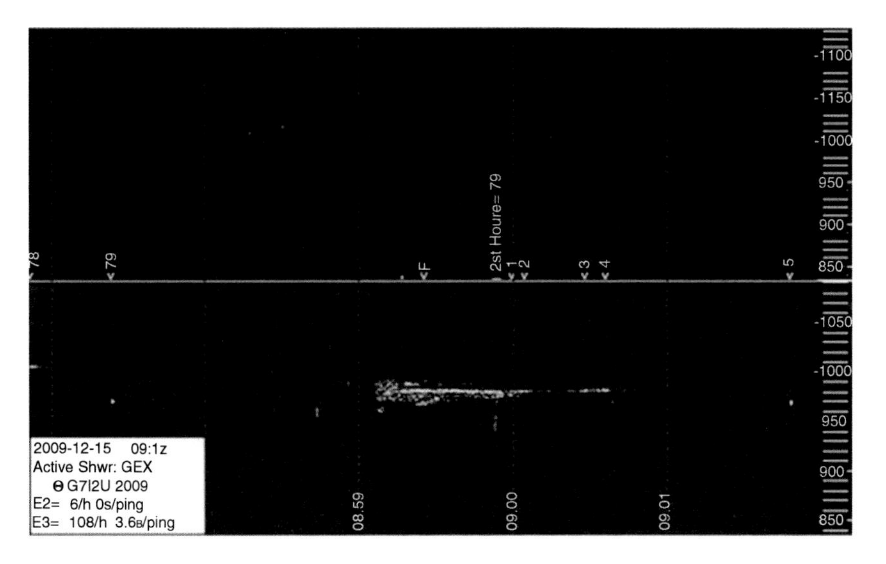

Fig. 13.9. Meteor radio spectrum of a Geminid taken on December 12, 2009, by Andy Smith G7IZU using Spectrum Lab software.

be configured to build up a count when a reflection event is detected, and if the reflection event was unusually long Spectrum Lab could be used to capture the spectrum. Long events could be due to fireball activity. This is the way Andy Smith uses Spectrum Lab to record meteor activity. An example of a spectrum he recorded is shown in Fig. 13.9.

Appendix A

Formulae in Radio Astronomy

Zenith Hourly Rate of a meteor stream calculated from observed visual rates:

$$ZHR = \left(\frac{N}{t}\right)\frac{r^{(6.5-L_m)}\left(\frac{1}{1-x}\right)}{\sin(\phi)}$$

where:

N is the observed number of meteors
t is the duration of the observing run in hours
L_m is the visual limiting magnitude
ϕ is the alt the altitude of the radiant in degrees
x is the percentage of sky obscured by cloud or obstruction
r is the population index of the stream, a property obtained from the stream data tables

Flux Density received by a radio telescope:

$$P = \frac{2\pi kT}{\lambda^2} \quad wm^{-2}Hz^{-1}$$

where:

S is the flux density
π is a constant 3.1415926
k is the Boltzmann constant 1.3806×10^{-23}
T is the temperature in kelvin
λ is the wavelength in meters
R is the radius of the object
d is the distance of the object (R and d in the same units).

The Gyrofrequency of an electron traveling in a magnetic field:

$$f = 2.8B$$

where f is the gyrofrequency in hertz and B is the magnetic field strength in gauss.

Tip to tip length of a dipole antenna:

$$L = \frac{143}{f}$$

where f is the frequency in megahertz and L is the total length of the dipole in meters.

Collecting area of a dipole antenna:

$$A = 0.13\lambda^2$$

where A is the area and λ is the wavelength, both expressed in the same units.

When the dipole is folded (total folded length found from the above formula), its bandwidth can be increased by placing a pair of shorting bars between the upper and lower parts of the folded loop. The distance of these from the feed point is found from:

$$L = \frac{61}{f}$$

where L is the length in meters, and f is the frequency in megahertz.

Quad loop beam antenna:

The total length of the driven (antenna) loop around its perimeter is found from

$$L_A = \frac{306.4}{f}$$

A slightly larger reflector is placed behind, and the reflector loop length is

$$L_R = \frac{315}{f}$$

and a slightly smaller director is placed in front whose length is

$$L_D = \frac{297}{f}$$

and the spacing between the elements is:

$$S = \frac{60}{f}$$

where L are the lengths in meters and f is the frequency in megahertz.

The shape of the loops can be round or square, but all three should match the same pattern.

Yagi Beam:

A three-element beam can be calculated from:

$$Director_{3e} = \frac{140.7}{f}$$

$$Dipole_{3e} = \frac{145.7}{f}$$

$$Reflector_{3e} = \frac{150}{f}$$

$$Spacing_{3e} = \frac{43.29}{f}$$

It is calculated in the same way as for a simple dipole.

And for a six-element beam using four directors all of the same dimensions and spacing:

$$Director_{6e} = \frac{134.39}{f}$$

$$Dipole_{6e} = \frac{145.05}{f}$$

$$Reflector_{6e} = \frac{145.56}{f}$$

$$Spacing_{6e} = \frac{44.8}{f}$$

All lengths are in meters, and f are in megahertz.

Collecting area of a Yagi antenna:

$$A = 0.65\lambda^2$$

where A is the area and λ is the wavelength, both in the same units.

Log periodic array:

The design of the log periodic is shown in Fig. 5.8. There are three parameters used to describe the dimensions, α, τ, and σ. The semi angle of the antenna is α. The other two are defined in the formulae below.

$$\tau = \frac{R_{n+1}}{R_n} = \frac{D_{n+1}}{D_n} = \frac{L_{n+1}}{L_n}$$

where all lengths R, D, and L are in the same units.

$$\sigma = \frac{1-\tau}{4tan(\alpha)} = \frac{D_n}{2L_n}$$

where α is in degrees, and the length units of D and L should be the same.
The longest element is calculated from the dipole formula:

$$L_{longest} = \frac{143}{f}$$

where L is in meters and frequency f in megahertz.

Beamwidth of a parabolic dish:

$$Beamwidth = \frac{57\lambda}{d}$$

where λ is the wavelength in meters and d the dish diameter in meters.

Effective wavelength of radio wave travelling in a circular waveguide:

$$\lambda_g = \frac{\lambda_0}{\sqrt{1-\left(\frac{\lambda_0}{\lambda_c}\right)^2}}$$

and

$$\lambda_c = 3.42r$$

where λ_g is the "guide wavelength," λ_0 is the wavelength of the signal in free space, λ_c is the cut-off wavelength, and r is the radius of the guide, keeping all the units of length the same.

The depth of a parabolic dish:

$$d = \frac{D^2}{16F}$$

where F is the focal length, D is diameter, and d is the depth of the dish from edge to center. All are in the same units.

Gain of a parabolic dish reflector:

$$G = 10log_{10}\left(k\left(\frac{D\pi}{\lambda}\right)^2\right)$$

where k is the efficiency of the dish (typically 0.55), D is the diameter, and λ is the wavelength. Gain G is in dB.

Length of time an object stays in a dish antenna beam:

$$T = \frac{13751\lambda}{Dcos(\delta)}$$

where λ is the wavelength, D is the diameter of the dish, T is in seconds, and δ is the object's declination.

Conversion of noise temperature to noise figure:

$$N_F = 10log_{10}\left(\frac{T+290}{290}\right)$$

where T is the temperature in kelvin, and N_F is the noise figure in decibel.

Bibliography

Radio Astronomy

Amateur Radio Astronomy, John Fielding. Radio Society of Great Britain (2006)

Although this book starts off with some early radio history, it continues with lots of useful fundamental information on the design and theory of radio and its application to astronomy. There are some good circuit diagrams included for project instruments. While this is included in the beginners category because it assumes no knowledge of radio astronomy, it is aimed at readers who are already involved in amateur radio, and therefore assumes a fair amount of radio electronics experience.

Radio Astronomy Projects, William Lonc. Radio Sky Publishing (1999)

A book based on a series of articles published in radio, physics, and astronomical journals. Not really a textbook but a description of a series of radio experiments. It is far more observationally oriented than construction based, with lots of ideas to base your own observational program on.

Radio Nature, Renato Romero. Radio Society of Great Britain (2008)

Not strictly speaking a book on radio astronomy, it covers topics of very low frequency radio study of Earth's environment such as lightning, whistlers, and aurorae. Some practical circuits and antenna designs are covered, too.

Radio Science Observing, Vol 1 & 2, Joseph J. Carr. Prompt Publications (1998, 1999)

These books are an excellent introduction to radio electronics, and they assume no prior knowledge. Many of the practical suggestions are geared up to low frequency operation similar to the book *Radio Nature.* However, it does describe hardware suitable for HF observation of Jupiter and the Sun. Lots of clear diagrams and practical circuit schematics included, but as with many such books it does not include PCB layouts. A pair of CD ROMs are included with some simple software applications.

Listen Up!, Laura A. Whitlock, Kiley Pulliam. iUniverse, Inc. (2008)

A series of laboratory exercises in radio astronomy with a small telescope. This is aimed directly at the user of "THE" Small Telescope (SRT), a kit-based instrument developed by the engineers at MIT Haystack Observatory and outlined on their website http://www.haystack.mit.edu/edu/undergrad/srt/index.html.

Radio Astronomy, John D. Kraus. Cygnus Quasar Books (1986)

Very technical book, lots of mathematics. It covers radio and radio astronomy theory followed by descriptive chapters on astronomical radio sources.

Tools of Radio Astronomy, Kristen Rohlfs. Springer (2008)

Like Kraus a very technical and mathematical coverage of radio astronomy and radio theory, followed by chapters on astrophysics pertaining to radio astronomy.

Meteor Science and Engineering, D. W. R. McKinley. McGraw-Hill (1961)

Published in 1961, but despite its age, it is still one of the most referenced books on the subject of radio meteor studies. It is still one of the best. Chapters briefly cover optical observation methods and a historical survey. Followed by chapters on the astrophysics of meteors and the theory of radio scattering, although aimed at radio communications it is still of considerable interest to the astronomer.

Astrophysics

Discovering the Secrets of the Sun, Rudolf Kippenhahn. John Wiley & Sons (1994)

This is a completely non-mathematical and highly descriptive coverage of the astrophysical processes occurring in the Sun. Some very difficult topics are described so very well it is highly recommended reading.

Space Science, Louise K. Hara, Keith O. Mason. Imperial College Press (2004)

This book covers the topics of planetary sciences, the Sun, space weather, space plasma, and space-borne hardware. It is a remarkably easy read on the whole, with minimal use of mathematics, although it gets heavy in parts.

Solar and Space Weather Radio Physics, Dale E. Gary, Christoph U. Keller. Kluwer Academic Publishers. (2004)

This is an expensive book, but is an excellent read when it comes to all things solar and space weather-related. It is highly recommended reading. There is lots to be learned from it about the radio properties of the Sun and the modern instrumentation used to study it.

Physics of the Jovian Magnetosphere, A.J. Dessler. Cambridge University Press (1983)

Published in 1983 this work is a little dated now, as it clearly does not cover the discoveries of the *Galileo* and *Cassini* spacecraft. However it still contains a lot of good material on the physics related to Jupiter's magnetosphere. It is of a highly technical nature, but despite that is very descriptive at times and does not rely quite so heavily on mathematics.

Astronomy Principles and Practice, A. E. Roy, D. Clarke, Institute of Physics (IOP) (2003)

A general textbook of observational principles, including sections on celestial mechanics and the nature of electromagnetic radiation. While aimed mostly at the optical observer there is a chapter on radio telescopes.

Astrophysics Processes: The Physics of Astronomical Phenomena, Hale Bradt, Cambridge University Press (2008)

This one is quite good at explaining, in depth, the astrophysical processes in many kinds of objects. Chapters are dedicated to thermal Bremsstrahlung, blackbodies, and synchrotron radiation, as well as Compton scattering, among others.

Astrophysics: Decoding the Cosmos, Judith A. Irwin, John Wiley & Sons (2007)

This is an excellent introduction to astrophysics and observational methods. It contains some chapters very relevant to radio astronomy and is fairly accessible, too, with some good descriptive sections not too heavily dependent on mathematics.

The Encyclopedia of Astronomy and Astrophysics. Nature Publishing, http://eaa.crcpress.com/subject to subscription

Although this used to be in the form of a four-volume printed encyclopedia, and it can still be purchased on the second hand market, it is now maintained and kept updated on the web. A subscription must be purchased. It is an invaluable reference to astrophysics on all topics, full of very good descriptive articles not too heavily dependent on mathematics.

Radio and Electronics

The ARRL Handbook, Editor Mark J. Wilson. The American Radio Relay League. (annual publication)

This is filled with information on radio theory and contains lots of practical circuit ideas for the amateur radio enthusiast. While aimed at amateur radio rather than amateur radio astronomy, it is still an extremely useful source of information on modern techniques of design and construction. A real must-have volume for the constructor.

The ARRL Antenna Handbook, R. Dean Straw, The American Radio Relay League (2007)

The companion volume to the above covering lots of antenna theory and practical projects, again aimed at amateur radio bands. Still a must-have volume for the constructor.

The RSGB Handbook (Radio Society of Great Britain, annual publication)

This is similar to the ARRL handbook, but published by the Radio Society of Great Britain. It, too, is a very useful read for anyone involved in radio construction.

The above handbooks are updated on a regular basis.

Antenna Toolkit, Jospeh J. Carr. Newnes (2001)

Another book by J.J. Carr, who always manages to explain radio theory in an easy-going and very clear style suitable for the beginner. Many different antenna types are explained, with design information included that allows you to adapt them to the frequency you are interested in.

Radio and Electronics Cookbook, George Brown. Radio Society of Great Britain/Newnes (2000)

Not much here for the radio astronomer, but this is a highly practical project book with clear instructions on building over 80 radio projects, The electronic novice will gain a lot of knowledge and experience building some of these units.

Secrets of RF Circuit Design, Jospeh J. Carr. TAB Electronics (2001)

You can't go wrong with a book by J.J. Carr. This book, like his others, provides excellent coverage of radio theory and design. Carr does not assume much prior knowledge on the subject.

RF Components and Circuits, Joseph J. Carr, Newnes (2002)

A good complementary book to the *Secrets of RF Circuit Design*, this one places an emphasis on how electronic components work at radio frequencies. This is an excellent reference to use when designing and constructing RF inductors based on toroidal cores.

RF Circuit Design, Chris Bowick, Newnes (2007)

It is not clear whether this book belongs in the advanced or beginner category. In many ways it is no more technical than the radio handbooks in the above section. However, it is more of a textbook than the others. It covers in detail an excellent method of designing filters. In many respects it is somewhat dated now, but it is of interest to constructors who may still want to build devices using old-style discrete components rather than surface mount chips.

To help you learn more about electronics, generally, especially microcontroller-based projects there are a few good monthly magazines, including are *Elektor*, *EPE*, and *Circuit Cellar.*

Books About Software

C Programming for Microcontrollers, Joe Pardue, Smiley Micros. (2005)

A good introduction to programming ATMEL microcontrollers using the free GNU GCC-based compilers, which includes lots of practical examples.

Processing: A Programming Handbook for Visual Designers and Artists, Casey Reas, Ben Fry, MIT Press. (2007)

Processing is an excellent language, is easy to learn, and has a lot of potential to produce highly visual PC applications that can interact with external hardware such as microcontrollers. Processing software is open source and free. It is available for Windows, Macintosh, and Linux.

Suppliers, Groups, and Societies

Equipment Plans, Kits, Parts, and Complete Setup Suppliers

There are not many places where you can just buy a radio telescope solution off the shelf, but here are a few places where design plans or even complete hardware units can be sourced.

Radio Astronomy Supplies

Run by Jeff Lichtman, this company can supply a range of simple to advanced instruments ready-built. Check out his website for details at http://www.radioastronomysupplies.com.

Radio Jove

A NASA-sponsored educational project. There are plans available and kits to build the Jove receiver, a 20.1 MHz solar and Jupiter radio telescope. The home page is http://radiojove.gsfc.nasa.gov.

Guru Receiver

The Guru receiver is essentially another version of the Jove receiver. Detailed descriptions and circuit diagrams can be downloaded from http://receiverforjupiter.tripod.com/receiver.htm.

INSPIRE

The INSPIRE project is all about studying natural radio at very low frequencies. Direct VLF is a way of studying Earth's environment. We have seen in this book that solar flares can also be detected at VLF. The INSPIRE project shows online details of their receiver, and kits can be ordered from http://theinspireproject.org.

The Small Radio Telescope (SRT)

The Small Radio Telescope was designed by a team at MIT at the Haystack Observatory. Details, plans, and software can be freely downloaded from their website at http://www.haystack.mit.edu/edu/undergrad/srt/index.html, and kits to build an SRT can be found at http://www.cassicorp.com. including a full x-y motorized dish solution.

The VSRT

This is another project from the MIT Haystack Observatory team. It's brand new, and as yet there does not appear to be a kit available, but information on its concepts can be downloaded at http://www.haystack.mit.edu/edu/undergrad/VSRT/index.html.

The eCallisto

Although this instrument is not available to buy as a kit, the plans for its construction, including software and PCB layouts, can be downloaded from the ETH Zurich website. The details can be found here http://helene.ethz.ch/instrument/callisto/ecallisto/applidocs.htm. The instrument is a VHF/UHF scanning spectrometer especially designed for studying solar activity.

UKRAA

UKRAA is a nonprofit organization founded by the British Astronomical Association Radio Astronomy Group in order to promote amateur radio astronomy. Its goal is to design and build low-cost hardware for a plug and play observatory solution, making it easier to simply observe and gather data for research projects. At the time of writing the organization had just been formed, and the first products available were a VLF receiver, a logger controller, and a magnetometer. Check out the website for information and contact details at http://ukraa.com.

Kuhne

Kuhne does not sell complete radio astronomy solutions or kits, but they sell lots of useful modules for building up a system, such as low noise amplifiers, down converters, etc. The site is in German but has English pages at http://www.kuhne-electronic.de/en

Minicircuits

Minicircuits is another non-specialized supplier of RF components and modules, such as amplifiers, mixers, oscillator's down converters, etc. Their home page is http://www.minicircuits.com.

Electronic Components, Tools, and Supplies

There are too many suppliers of parts to mention them all. Many specialize in particular parts or market areas. The following are some of the general-purpose suppliers of parts, materials, and tools. They carry large ranges and stocks and can supply worldwide in most cases.

Farnell	http://www.farnell.com
RS Components	http://uk.rs-online.com
Digi-Key	http://www.digikey.com
Rapid Electronics	http://www.rapidonline.com
Conrad	http://www.conrad.com

eBay

One of the best places to buy all kinds of items, there is probably an eBay site dedicated to most countries, and many traders sell worldwide, too. eBay is not always the cheapest place to buy, but sometimes you can find components you can't easily find elsewhere, such as obsolete but still useful electronic components.

Groups and Societies

British Astronomical Association Radio Astronomy Group	http://www.britastro.org/radio
SARA	http://www.radio-astronomy.org
SETI League	http://www.setileague.org
ERAC	http://www.eracnet.org
ARAE	http://arae.ukaranet.org.uk
International Meteor Organization	http://www.imo.net
Radio Meteor Observers Bulletin (RMOB)	http://www.rmob.org

Appendix D

Glossary

Absolute magnitude	The magnitude a star as it would appear at a distance of 10 pc.
AM	Amplitude modulation, where the amplitude of a radio wave carries audio information.
Amplifier	A device used to increase the amplitude of a signal.
Amplitude	A measure of the peak value of an oscillating waveform.
Astronomical unit (AU)	Mean Earth-to-Sun distance, about 150,000,000 km.
Azimuth	The horizontal angle measured from the north to the point immediately below an object on the horizon.
Background radiation	Electromagnetic radiation originating from no specific location.
Beamwidth	The angle within which an antenna receives radio waves.
Blackbody	A hypothetical concept of a perfect radiator, absorbing and reemitting all radiation that impinges on it.
Black hole	A region of space surrounding a massive collapsed star from which lisght can't escape.
Bowshock	The boundary of a planet's magnetosphere facing the Sun.
Brightness	Power of radiation per unit area per unit solid angle per unit hertz.
Brightness spectrum	Variation of radiation brightness with frequency.
Chromosphere	The layer of the Sun's atmosphere just above the photosphere.
Corona	The outer atmosphere of the Sun.
Cyclotron radiation	EM radiation emitted when charged particles are moving in a magnetic field at non-relativistic speeds.
Decametric emission	Radio emission in the tens of meters wavelength range.
Decimetric emission	Radio emission in the tens of centimeters wavelength range.
Declination	Angular distance north or south of the equator of an object in the sky.

Detector	A device used to recover information from a waveform.
Diffraction	The bending of electromagnetic waves as they pass by obstructions.
Discrete source	A source of radiation whose direction can be identified.
Doppler effect	Apparent change in wavelength of the radiation from a source due to its motion relative to the observer.
Down converter	The combination of an oscillator and mixer intended to reduce the frequency of a signal.
Electromagnetic radiation	Radiation consisting of a combination of an electric and a magnetic wave.
Electromagnetic spectrum	The full range of wavelengths or frequencies of EM radiation.
Elevation	In the horizon coordinate system, the angle upward from the horizon to an object in the sky.
Emission lines	Bright lines in a spectrum.
Equator	A great circle on Earth 90° from the poles.
Extended source	Discrete emitter of radiation that covers a relatively large area.
Faraday rotation	Rotation of an electromagnetic wave's polarization as it passes through a magnetic field parallel to the propagation of the wave.
Field	The effect of forces, such as electric or magnetic, that act on objects.
Filter	A device used to restrict the range of frequencies which can pass.
FM	Frequency modulation. Where audio information is transmitted by varying the frequency of a wave.
Hertz	The unit of measure of frequency, in cycles per second.
Horizon coordinate system	A coordinate system using a plane through the observer parallel to the horizon as a reference.
Hour angle	The elapsed time since an object in the sky crossed the meridian.
Hour circle	A great circle on the celestial sphere passing through both poles and a celestial object.
IF	Intermediate frequency; the frequency output from a mixer.
Index of refraction	Ratio of the speed of electromagnetic radiation in a vacuum to its speed in a given medium.
Intensity	Power of electromagnetic radiation received per unit area.
Interference	The additive combination of two electromagnetic waves.
Interferometry	In radio astronomy, use of more than one radio telescope or antenna to enhance the resolution of a radio telescope.

Inverse square law	The amount of electromagnetic energy reaching a given point is reduced by the square of the distance from the source.
ISM	Interstellar medium.
Kelvin	A unit of absolute temperature.
ku band	Range of radio frequencies of about 11–18 GHz.
L band	Range of radio frequencies of about 1–2 GHz.
LNB	Low noise block; the combination of filter, amplifier, oscillator, and mixer, whose output frequency is lower than the input frequency.
Localized source	A discrete radiation source of very small extent.
Magnetosphere	Zone around a planet in which its magnetic field dominates over the interplanetary environment.
Maser	Microwave-amplified stimulated emission of radiation. The radio equivalent of a laser.
Meridian	A great circle on the celestial sphere that passes through the poles and the zenith.
Microwave	A range of frequencies above about 2 GHz.
Mixer	A device designed to alter the frequency of an RF signal, usually used to reduce it to a lower, more manageable, frequency.
Nadir	The point on the celestial sphere opposite the zenith.
Nebula	Cloud of interstellar gas or dust.
Neutron star	A star of extremely high density, composed almost entirely of neutrons.
Non-thermal emissions	Electromagnetic radiation-produced mechanisms not related to temperature.
Oblateness	The flattening of a sphere.
Optical window	The characteristic of Earth's atmosphere that allows visible light to pass through.
Oscillator	A device or circuit used to generate an oscillating electrical waveform.
Parsec	The distance of an object at which 1 AU would subtend 1 arc second.
Phase	Angular distance between peaks or troughs of two waveforms of similar frequency.
Photon	A packet of electromagnetic energy; the particle form of an EM wave.
Photosphere	The region of a star from which continuum spectral radiation escapes into space, where the gases turn opaque.
Plasma	Hot ionized gas.
Point source	An idealized discrete source of radiation that subtends an infinitesimally small angle.
Polarization	The direction of the electric vector of an electromagnetic wave.
Preamplifier	A wideband amplifier whose input is unfiltered.

Preselector	An amplifier whose input range is restricted by a filter.
Pulsar	A neutron star radio source that emits rapid regular pulses.
Quasar	A very distant very luminous source of EM radiation.
Radio window	The property of Earth's atmosphere that allows certain wavelengths of radio to pass through.
Red shift	Apparent lengthening of the wavelength of radiation received from a source caused by the Doppler effect.
Rj	Abbreviation for Jupiter's radius (about 5 million km).
S-band	Range of radio frequencies of about 2–4 GHz.
Scintillation	Effect produced by phase shifting of electromagnetic waves from a discrete source as they pass through Earth's ionosphere, or other ionized plasma.
SETI	Describes several programs involving the search for extraterrestrial intelligence.
Sidereal day	The time required for Earth to revolve once on its axis with respect to a distant celestial object.
Solar day	The time required for Earth to revolve once on its axis with respect to the Sun.
Solar flare	Brilliant outbreak in the Sun's outer atmosphere, usually associated with active groups of sunspots.
Spectral power	Power of electromagnetic radiation per unit of frequency bandwidth.
Stellar wind	The outflow of gas from a star.
Sunspot	A temporary cool region in the solar photosphere that appears dark by contrast.
Supernova	A star that experiences a cataclysmic explosion, breaking up the star and resulting usually in a black hole or neutron star.
Superposition	The ability of electromagnetic waves to traverse the same space independently of each other.
Synchrotron radiation	Radiation emitted by relativistic charged particles accelerated in a magnetic field.
Thermal emissions	Radiation emitted due to an object's temperature.
X-band	Range of radio frequencies of about 8–12 GHz.
Zenith	The point on the celestial sphere directly overhead.

Index

Index

Manufactured by Amazon.ca
Bolton, ON